WE ARE NOT THE FIRST

WE ARE NOT THE FIRST

Riddles of Ancient Science

ANDREW TOMAS

G. P. Putnam's Sons
New York

Contents

	Introduction	13
1	The Days and Nights of Knowledge	18
2	Novelties in Antiquity	31
3	Discoveries Create Problems	39
4	The Blacksmith of Olympus	48
5	The Forgotten Art of Goldmaking	52
6	The Caduceus of Hermes	67
7	From Temples and Forums to Atomic Reactors	72
8	Sages Under the Heavenly Vault	78
9	The Zodiac and the Music of the Spheres	93
10	Apes and Ages	97
11	The Celestial Comedy	104
12	Maps, Manuscripts, and Marvels	109
13	Electricity in the Remote Past	119
14	Did the Ancients Master Gravitation?	125
15	Prehistoric Aircraft	136
16	They Conquered Space Long Before We Did	142
17	First Robots, Computers, Radio, Television, and Time-Viewing Machines	153
18	An Enigma of Social Science—the Incas	167
19	Apollonius Met the Men Who Knew Everything	175
20	Diamonds and Stars; The Immortal Saint-Germain	179
21	In the Abode of Wisdom—Roerich	193
22	In Quest of the Source	198
	Conclusion	215
	Rediscovery of Science	217
	Bibliography	219

7

Acknowledgment

The author would like to express his appreciation of the assistance rendered by Elaine Ackerman and Anne Croser, both of Paris, in checking the manuscript and offering constructive criticisms.

Illustrations

Follow page 112.

Prehistoric rock painting near Brandberg, South-West Africa

Babylonian priest-astronomers

A mysterious rock carving near Navai, Uzbekistan, USSR

The enigma of the round hole in an ancient auroch's skull

Alchemically transmuted lower portion of a silver medal into gold

Dragon-chariot in the clouds

Admiral Piri Reis' enigmatic maps dated 1513 and 1528

Ancient India had notions of aviation as this bas-relief suggests

Magnetic Hill, New Brunswick, Canada, where cars go uphill without power

Rock carving near Alice Springs, Australia

Ancient cave painting in the Prince Regent River Valley, Australia

Fragments of an ancient Greek mechanical device

The Antikythera computer

Blocking the Gulf Stream, Atlantis might have been the cause of the last Ice Age

The sinking of Atlantis could have ended the Ice Age by letting the warm Gulf Stream advance northward

9

Quetzalcoatl, the Feathered Serpent or the Morning
Star

Soviet science fiction writer Kazantsev sees a jet rocket
in this Mayan carving

The heavy boulder in front of a Mohammedan shrine
in Shivapur, India, which becomes weightless

The Comte de Saint-Germain

Apollonius of Tyana

Oannes, the civilizing influence of the Babylonians
and Sumerians

Nicholas Roerich in Mongolia

An ancient Greek philosopher?

For the illustrations of the Antikythera computer the author wishes to thank the National Archaeological Museum of Greece, Dr. Derek J. de Solla Price of Yale University, and the *Scientific American*. The photograph of Wenzel Seiler's medallion is reproduced by permission of the Kunsthistorisches Museum of Vienna. The author is grateful to Magnetic Hill Enterprises Ltd. of Magnetic Hill, New Brunswick, Canada, for the photograph of Magnetic Hill, and to the South African Tourist Board for the photograph of the prehistoric rock painting from South-West Africa.

WE ARE NOT THE FIRST

Introduction

Hundreds of intellects, past and present, played a part in this book. The author acted merely as an orchestra conductor. His musicians comprised classic writers; priests of old Egypt, Babylon, India, and Mexico; philosophers of ancient Greece and China; scholars of the Middle Ages, and lastly modern scientists. The theme of this composition is the Genesis of Knowledge and its periodic crescendos and diminuendos in history.

Three aims are set in this work:

—To show that in former eras people possessed many scientific notions that we have today.
—To demonstrate that the technical skills of the men of antiquity and prehistory have been greatly underestimated.
—To prove that certain advanced ideas of the ancients on science and technology came from an unknown outside source.

"Civilization is older than we suppose" is the principal thesis of this treatise.

With the advance of science the concept of the size and age of the universe has been radically changed in the last four hundred years. Farseeing men such as Bruno, Galileo, or Darwin defied their narrow-minded contemporaries and argued that the world was greater and more ancient than men had believed.

Two hundred years ago the French naturalist Buffon estimated the age of the earth. He thought that our planet had cooled down 35,000 years ago, and that life had appeared about 15,000 years ago. This chronology of the French scholar was certainly more rational than the general belief in England at the time of the coronation of Queen Victoria in 1837 that the earth and man were created in 4004 B.C.

But geology and Darwinism exploded this medieval concept and twenty-five years later Lord Kelvin added 10 million years to the terrestrial age. Thanks to perfected techniques the age of the crust has been determined to be 3.3 milliard years while that of the planet itself—4.6 milliard years. In only two hundred years the age of the earth's crust was raised from 35,000 to 3,300,000,000 years!

A few decades ago man was considered to be about 600,000 years old. New finds in South and East Africa extended the life-span of *Homo sapiens* to two million years. The discovery of anthropoid teeth and jaws in southern Ethiopia by the Chicago anthropologist F. Clark Howell in 1969 confirms the figure.

A tendency to move back the origin of civilization has been noticeable in the field of history as well. Before Schliemann no savant in Europe could conceive of Troy's existing as early as 2800 B.C. Before the excavations of Evans in Crete no historian had the audacity to imagine a Cretan culture 2,500 years before our era. Four decades ago there was not a scholar in the world who could envisage a high civilization in the Indus Valley coexisting with the early dynasties of Egypt. How many scientists were there a quarter of a century ago who could accept the idea of Central American civilizations having had an uninterrupted existence for 4,000 years? Yet the ruins of the city of Dzibilchatun in Yucatán are a mute witness to this truth.

The rationale of the conclusion that the origin of man and the appearance of civilization might be less recent than accepted at present can clearly be seen from the above examples.

The mass of historical data presented in this work demonstrates the presence of an archaic science in the past. But who were the teachers of the ancient Egyptians, Babylonians, and Greeks, from whom we ourselves received a store of knowledge through the Arabs?

Surrounded by the marvels of our technology and science, we are losing contact with the people of former epochs to whom we owe so much.

Man is civilized only when he remembers his yesterday and dreams about tomorrow. The primate began to leave the animal kingdom by developing a superior brain and upright posture. He became a true man when he ventured into the domain of abstract thought—religion, mathematics, art, and music.

The true criterion of man's growth was his ability to soar in the world of ideals—to appreciate beauty, to distinguish between right and wrong, to form abstractions. Until he reached that state, man was still a link between the quadrupeds and the bipeds.

Science, the empirical observation of the world around us, and philosophy, the formulation of generalizations, have helped man to arrive at more correct views concerning the universe.

The history of civilization is the story of the ascent of man in the mental world. It was William Prescott, the great Americanologist, who said, "A nation may pass away and leave only the memory of its existence, but the stories of science it has gathered up will endure forever."

Have you, like the author, walked around the pyramids of Giza awed by the giant size of the stones and amazed by the thinness of the joints between them?

Have you stood in Mexico City before the paintings of Quetzalcoatl flying in a winged ship, startled by this prehistoric notion of aviation?

Have you seen canoes with stabilizers in the Pacific and admired the brown islanders who have made ocean voyages for thousands of years?

Have you strolled through the slumbering city of Pompeii and examined the bricks with incrusted gravel, like modern concrete, which the Roman slaves made?

Have you visited the shrine inside the colossal bronze Buddha in Kamakura, Japan, and marveled at the skill of Japanese metallurgists 700 years ago?

Have you walked around the megaliths of Stonehenge and tried to solve a riddle—how men wearing skins could have designed and erected this computer in stone?

If you have, then you would want to follow the author on a sight-seeing trip to the land of the past. This book is about real people, actual places, and authentic events. What is more, it is about the things our predecessors thought and dreamed about a long time ago.

During the past three or four hundred years science has been rediscovered rather than discovered. Babylon, India, Egypt, Greece, and China were the cradle of science. "The old devices have been re-invented; the old experiments have been tried once more," said Alexander Graham Bell, the inventor of the telephone.

This book is about penicillin before Fleming, about airplanes before the Wright brothers, about the moons of Jupiter before Galileo, about voyages to the moon before Apollo probes, about the atomic theory centuries before Rutherford, about electric batteries before Volta, about computers before Wiener, about science before this Science Age.

A fragmentary account of the adventures of ancient man in the scientific realm is not a history of science. But this sketch will reveal unorthodox historical facts of educational value, provoke speculation as to the reasons for the presence of advanced scientific and technological concepts in early civilizations, or at least entertain the reader by a story stranger than fiction.

1

The Days and Nights of Knowledge

The world is rectangular, stretching from Iberia (Spain) to India, and from Africa to Scythia (Russia). Its four sides are formed by high mountains on which rests the celestial vault. The earth is but a chest of giant dimensions, and on the flat bottom of this coffer are all the seas and lands known to man. The sky is the lid of this trunk and the mountains are its walls. This is the childish image of the earth drawn by Cosmas Indicopleustes, a scholar-explorer of the sixth century, in his *Christian Topography*.

But a thousand years before Cosmas' book, philosophers had a different and much more accurate idea of the shape of the earth. Pythagoras (sixth century B.C.) taught at his school in Crotona that the earth was a sphere. Aristarchus of Samos (third century B.C.) deduced that it revolved around the sun. Eratosthenes, the librarian of Alexandria (third century B.C.), computed the circumference of our planet.

Oddly enough, the peoples further back in time had greater scientific knowledge than the nations of later historical periods. Until the second part of the nineteenth century scholars and clerics of the West thought that the earth was but a few thousand years old. Yet ancient Brahmin books estimated the Day of Brahma, the life-span of our universe, to be 4.32 billion years. This figure is close to that of our astronomers, who calculate it to be about 4.6 billion years.

It is fairly obvious that knowledge has had its days and nights. Science emerged from medieval darkness during the Renaissance. By studying classic sources savants rediscovered truths which had been known to ancient Babylonians, Egyptians, Hindus, or Greeks for many centuries.

These waves of progress can be followed back over seven or eight thousand years—the frontiers of history. They can be explained by changes in ideology or by a new economic or political system, as well as by the impact of great minds on society. However, the presence of certain scientific knowledge in ancient times cannot be easily accounted for unless it is assumed that the skills and knowledge of the ancients have been drastically underestimated. Even then a certain number of riddles will remain, calling for a reappraisal of the history of science. The presentation of these problems constitutes one of the aims of this book.

In 1600 a Dominican monk, Giordano Bruno, was burned alive at the Piazza del Fiore in Rome after having been convicted as a heretic. In one of his books he stated that there are an infinite number of suns in the universe and planets which revolve around them. Some of these worlds might be populated, he said.

This brilliant speculation of Bruno, though 400 years ahead of his era, actually preceded him by 2,000 years, because ancient Greek philosophers had believed in the plurality of inhabited worlds. Anaximenes told disappointed Alexander the Great that he had conquered but one earth whereas there were many others in infinite space. In the third century B.C. Metrodotus did not think that our earth was the only populated planet. Anaxagoras (fifth century B.C.) wrote about "other earths" in the universe.

Until Descartes and Leibniz the Europeans had no concept of the million in mathematics. But the ancient Hindus, Babylonians, and Egyptians had hieroglyphs for one million, and manipulated astronomical figures in their records. The Egyptians had an apt symbol for the million—an amazed man with raised hands. We owe everything in mathematics and science to ancient India for the most important and yet cheapest gift to the world—the zero.

Medieval cities of France, Germany, England, and other countries were usually built by accident without any planning. The streets were narrow, irregular, with no sewage facilities. Because of unsanitary conditions epidemics devastated these crowded towns.

But around 2500 B.C. the cities of Mohenjo Daro and Harappa, in what is now Pakistan, were as carefully planned as Paris or Washington. Efficient water supply, drainage, and rubbish chutes were provided. Besides public swimming pools many homes had private bathrooms. Let us recollect that until the end of the last century this was a luxury in Europe and America.

Before the latter part of the sixteenth century Europeans had neither spoons nor forks on their tables—they used only knives and fingers. Yet the people of Central America had spoons and forks 1,000 years before the appearance of Cortés. In fact, the ancient Egyptians had used spoons even earlier—in 3000 B.C. This historical detail puts things in the proper perspective to the shame of the Europeans.

The Aztecs were living in a Golden Age when the conquistadors invaded Mexico—Moctezuma actually walked on gold because his sandals had soft-gold soles. There was also a Golden Age in the land of the Incas when the Spaniards came—the temples of Pachacamak near Lima were fastened with golden nails that were found to weigh a ton, when removed. There was a Silver Age in Peru in the days of Pizarro—his soldiers shod their horses with silver.

To show how the expansion of Europe was conducted at the expense of the Golden Age races of the Americas, let us examine the gold reserves of European nations in the year 1492, when Columbus set sail on his voyage to the New World. The total amount of gold in Europe at the time was ninety tons. After the robbing of the empires of Mexico and Peru, the gold reserves of Europe had increased eight times just one hundred years later!

But was there a Golden Age of Science? Did the priests of Peru, Mexico, India, Egypt, Babylon, and China and the thinkers of Greece endeavor to preserve its memory?

Our science has only rediscovered and perfected old ideas. Step by step it has demonstrated that the world is more ancient and vaster than it was thought to be only a few generations ago. In the last 150 years the space-time frontiers of the universe have been pushed far back.

In the fluctuations of scientific knowledge down through the ages a curious fact becomes evident—the possession of information which could not have been obtained without instrumentation. Occasionally knowledge has appeared as if out of nowhere. These problems require an unbiased approach.

One of the greatest handicaps that the historian is confronted with is lack of evidence. If it were not for the burning of libraries in antiquity, history would not have had so many missing pages. The past of many a former civilization would look different without these blank spots.

Firstly, let us review this destruction of cultural records. The famous collection of Pisistratus in Athens (sixth century B.C.) was ravaged. Fortunately the poems of Homer, edited by the Greek leader's literati, somehow survived. The papyri of the library of the Temple of Ptah in Memphis were totally destroyed. The same fate befell the library of Pergamus in

Asia Minor containing 200,000 volumes. The city of Carthage, razed to the ground by the Romans in a seventeen-day fire in 146 B.C., was said to possess a library with half a million volumes. But the greatest blow to history was the burning of the Alexandrian library in the Egyptian campaign of Julius Caesar during which 700,000 priceless scrolls were irretrievably lost. The Bruchion contained 400,000 books and the Serapeum 300,000. There was a complete catalogue of authors in 120 volumes with a brief biography of each author.

The Alexandrian library was also a university and research institute. The university had faculties of medicine, mathematics, astronomy, and literature, as well as other subjects. A chemical laboratory, an astronomical observatory, an anatomical theater for operations and dissections, and a botanical and zoological garden were some of the facilities of this educational institution, where 14,000 pupils studied, laying the foundation of modern science.

The Roman conqueror was also responsible for the loss of thousands of scrolls in the Bibracte druid college at what is now Autun, France. Numerous treatises on philosophy, medicine, astronomy, and other sciences perished there.

The fate of libraries was no better in Asia, as Emperor Tsin Shi Hwang-ti issued an edict whereby innumerable books were burned in China in 213 B.C.

Leo Isaurus was another archenemy of culture as 300,000 books went to the incendiary in Constantinople in the eighth century. The number of manuscripts annihilated by the Inquisition at *autos-da-fé* in the Middle Ages can hardly be estimated.

Because of these tragedies we have to depend on disconnected fragments, casual passages, and meager accounts. Our distant past is a vacuum filled at random with tablets, parchments, statues, paintings, and various artifacts. The history of science would appear totally different were the book   collection   of   Alexandria   intact   today.

Losses of priceless documents have occurred in modern history. Once there was a fire in the harem of the sultan of the Ottoman Empire. A young secretary of the French Embassy drifted with the crowds nearby and saw looters carry vases, curtains, and other objects from the burning palace. The Frenchman noticed a man with a thick volume of Titus Livius' *History of Rome,* considered lost for centuries. He immediately stopped the Turk, offering him a goodly sum for the book. Unfortunately, his purse contained only a few coins and he promised to pay the balance at his residence, to which the Turk agreed. Suddenly, they were separated by the mob before introducing themselves. This is how an irreplaceable document vanished after nearly having been retrieved.

On the other hand, unexpected discoveries which filled gaps in ancient history have been made. About one hundred and fifty years ago the great French Egyptologist Champollion visited the Turin Museum. In a storeroom he came across a box with pieces of papyri.

"What's in it?" he asked.

"Useless rubbish, sir," answered the attendant.

Champollion was not satisfied with the answer and began to put the pieces together like a jigsaw puzzle. This "rubbish" turned out to be the only extant list of Egyptian dynasties with the names of the pharaohs and the dates of their reigns! It was a revelation. One can imagine how our views on antiquity will change if more chronicles of this type are found—but not in rubbish bins.

The sensational find of the Dead Sea Scrolls disclosed the fact that this older version of the Bible (second century B.C.) agreed reasonably well with the Masoretic Text (tenth century A.D.). From an historical and religious standpoint the Dead Sea Scrolls were a tremendously important acquisition. Apropos, the credit for this momentous archaeological discovery is attributed to a young Bedouin shepherd who one day, while chasing a goat, found the cave in which the scrolls were hidden in jars.

In 1549 a young, overzealous monk, Diego de Landa, discovered a large library of Mayan codices in Mexico. "We burned them all because they contained nothing except superstition and machinations of the devil," he wrote.

How could he possibly know what the books were about? With all the brilliant philologists and electronic brains we have today, the three miraculously surviving manuscripts of the Mayas still remain undeciphered.

When de Landa had become older and received the title of bishop, he realized what a barbarian crime he had committed. He made a search for Mayan scripts but without success. A tradition exists that the fifty-two golden tablets preserved in a temple, containing a history of Central America, had been carefully concealed by the Aztec priests before the greedy conquistadors reached Tenochtitlán.

Diego de Landa wrote a work on the Mayas but his contribution toward decoding their hieroglyphics was utterly negligible.

Had someone requested the Madrid library 100 years ago for the *First New Chronicle and Good Government,* by Felipe Huaman Poma de Ayala, with the date of 1565, the librarian would have been extremely puzzled. Neither the Madrid librarian nor any scholar in the world at that time knew anything about this history of the Incas. The manuscript lay in obscurity for centuries until it turned up, of all places, in the Royal Library at Copenhagen in 1908. It was published for the first time in 1927, and is now considered to be as good a source as Garcilaso de la Vega or Pedro de Cieza de León. This is the story of just one lost book, but how many others may still be concealed in the most unexpected places?

Until these documents from bygone epochs are located, can the only sacred texts, classic writings, and myths we know of now be considered as reliable material for reconstructing the picture of the past? Sacred scriptures as well as the works of Greek and Roman authors can safely be used for this purpose. This contention will be supported later by interesting episodes from ancient history. Mythology and folklore are thought-fossils depicting the story of vanished cultures in symbols and allegories. By the separation of fancy from fact, a reasonably correct image of past events, people, and places can be re-created from legends.

The city of Ur, referred to in the Bible as the town from which Abraham had come, was not afforded any geographical or historical significance by the sages of the nineteenth century. Actually, until recent times few historians have taken the Bible seriously as a source of historical data. But after Sir Leonard Woolley had discovered the ancient city of Ur in Mesopotamia, the situation began to change.

One hundred years ago no scholar took the *Iliad* or the *Odyssey* of Homer as history. But Heinrich Schliemann put faith in it and discovered the legendary city of Troy. Then he followed the homeward route of Odysseus and excavated the graves of Mycenae in search of the loot which the Greeks took from Troy. He read in the *Iliad* Homer's description of a cup decorated with doves which Odysseus used. In a deep shaft Schliemann found that 3,600-year-old cup!

Legends can therefore be interpreted as fanciful records of actual happenings. Thus, for instance, the legend of the goddess Demeter, who is usually depicted with a sickle and sheaves of wheat, portrays the introduction of wheat into Greece, where up to that time there had been only beans, poppy seeds, and acorns. The goddess taught Triptolemus the art of agriculture and then he traveled throughout Greece instructing the people how to grow wheat and bake bread.

The myth of the birth of Zeus in Crete points to the Cretan origin of the ancient Greek culture. It is interesting to note that with the exception of a few legends, the Greeks themselves knew nothing of the advanced Minoan civilization in Crete, which had preceded their own. But, as we can see, folklore preserves history in the guise of colorful tales.

Until 1952, when Michael Ventris decoded the Linear B script of Crete and ascertained to his amazement that it was early Greek, no one in ancient or modern times had taken this Zeus myth seriously.

In his *Dialogues,* Plato made a reference to an archaic form of the Greek language. Naturally, his contemporaries had never heard of this lost dialect. But late in the nineteenth century an old script was found which, when deciphered in the fifties, turned out to be preclassic Greek. Consequently, do we have the right to distrust the words of antique writers or legends, unless and until they can be proved wrong?

In the *Critias* Plato tells the story of Solon, to whom the priests of Saïs in Egypt confided in 550 B.C. that 9,000 years before their time Greece had been covered with fertile soil. "In comparison to what was then, there remain in small islets only the bones of the wasted body, as they may be called, all the richer and softer parts of the soil having fallen away," said the Egyptian sages.

Now that information is scientifically correct because the soil of Greece was rich a few thousand years ago. In that remote period the Sahara was a steppe where abundant vegetation grew. This is but one example of the climatic changes which have taken place in the Mediterranean basin. But how could Plato, Solon, or the priests of Saïs have known about soil erosion in Greece for so long a period unless accurate records had been kept for 10,000 years by the Egyptian priesthood?

In describing the far north of Scythia, or Russia, Plutarch (first century A.D.) spoke of a night which prevailed for six months in those regions, with a continuous snowfall. He remarked that "this is utterly incredible." But his portrayal of the arctic winter was surprisingly true.

26

Plutarch also wrote a story concerning a Phoenician fleet in the service of Pharaoh Necho. The ships sailed from the Red Sea into the Indian Ocean and circled Africa via the Cape of Good Hope and the Straits of Gibraltar. The voyage was accomplished in the course of two years.

"These men made a statement," writes Plutarch, "which I do not myself believe, though others may, to the effect that as they sailed on a westerly course around the southern end of Africa, they had the sun on their right—to northward of them." No ancient Greek could imagine the sun shining in the north. The critical attitude of Plutarch to his own story adds even more weight to his writings. After all, his report is accurate as the sun does shine in the north in South Africa.

Ptolemy's *Almagest** lists all the geographical data available in the second century A.D. The astronomer describes equatorial Africa, the Upper Nile, and mountain ranges in the heart of the continent. Clearly, this savant of antiquity had more knowledge about Africa than his European colleagues in the first half of the nineteenth century.

When the exploration of central Africa in the last century disclosed the existence of snow-capped ranges, and reports to that effect were submitted to the Royal Geographical Society in London, the learned members found them a source of merriment. Snow on the equator? Sheer nonsense! The weapon of skepticism is dangerous—in the past many an overskeptical scientist discredited himself by rash condemnation and lack of imagination.

In explaining the causes of the flooding of the Nile, Herodotus (fifth century B.C.) listed several theories current at the time. One of them, "the most plausible" in his words, but impossible in his opinion, was that "the water of the Nile comes from melting snow."

So once again it is seen how the curve of knowledge plunges on the chart of world progress. It is not difficult to prove the

*_Lunae Mons,_ Geographia IV, 8.

superiority of Greek thought over scholastic philosophy in the Dark Ages. Born in antiquity, eclipsed in the medieval era, science was rediscovered by the Arabs, restored in the Renaissance, and developed by the scientific men of modern times.

But long ago there were other ebbs and flows of cultural progress. The rock paintings of aurochs, horses, stags, and other beasts in the caves of Altamira, Lascaux, Ribadasella, and others are masterpieces not only of prehistoric art but of art in any period.

Ancient Egyptians, Babylonians, and Greeks painted stylized bulls. But the bisons or horses of Altamira or Lascaux look as if they might have been painted by a Leonardo or a Picasso. The realism and beauty of these cave paintings makes them immensely superior to the paintings of animals in Egypt, Babylon, or Greece.

Sketches and trial-pieces have been discovered in the caves, suggesting the existence of art schools over 15,000 years ago. The rock paintings of the Cro-Magnon are more than 10,000 years older than the artistic productions of ancient civilizations. This is yet another example of the way a wave reaches a peak in the curve of civilization and then goes down.

Recently we have been rediscovering a forgotten science. Three hundred and fifty years ago the great German astronomer Johann Kepler correctly attributed the cause of tides to the influence of the moon. However, he immediately became a target for persecution. Yet, as early as the second century B.C., the Babylonian astronomer Seleucus spoke about the attraction which the moon exercises on our oceans. Posidonius (135-50 B.C.) made a study of the tides and rightly concluded that they were connected with the revolution of the moon around the earth. The eclipse of science in the intervening eighteen centuries is only too obvious.

During the course of fourteen centuries—from Ptolemy to Copernicus—not a single contribution to astronomy was made. Even in Ptolemy's time thinkers looked back to former centuries for knowledge as if there had been a Golden Age of Science in the past.

The ancient Indian astronomical text *Surya Siddhanta* recorded that the earth is "a globe in space." In the book *Huang Ti-Ping King Su Wen* the learned Chi-Po tells the Yellow Emperor (2697-2597 B.C.) that "the earth floats in space." Only four hundred years ago Galileo was condemned by the ecclesiastical authorities for teaching this very concept.

Diogenes of Apollonia (fifth century B.C.) affirmed that meteors "move in space and frequently fall to the earth." Yet the eighteenth-century pillar of science Lavoisier thought otherwise: "It is impossible for stones to fall from the sky because there are no stones in the sky." We know now who was right.

Twenty-five hundred years ago the great philosopher Democritus said that the Milky Way "consists of very small stars, closely huddled together." In the eighteenth century the English astronomer Ferguson wrote that the Milky Way "was formerly thought to be owing to a vast number of very small stars therein; but the telescope shows it to be quite otherwise." Without a telescope Democritus was certainly a better astronomer than Ferguson. It was a case of "a large telescope but a small mind" against "a great mind without a telescope."

When Marco Polo, his father, and his uncle returned to Venice from the Far East in their dusty, outlandish caftans, they were not recognized at first. Because of Marco Polo's stories of the fabulous riches of China and Japan, he was immediately nicknamed *Messer Millione* or "Mister Million." A dinner served by the Polos' relatives was attended by the notables of Venice. Then unexpectedly the Polos cut the linings of their heavy coats. Cascades of precious gems flowed onto the table! The Venetians could only gasp—Marco was telling the truth after all. There were rich empires in the Far East—these diamonds, rubies, sapphires, jades, and emeralds were a spectacular corroboration of his adventures!

The next chapter presents a number of curiosities from the history of science. They are like the jewels of Marco Polo—tangible evidence of a distant source of science.

2

Novelties in Antiquity

The achievements of modern science are phenomenal but with our background of spaceships, skyscrapers, wonder drugs, and atomic reactors we are apt to minimize the scientific accomplishments of the ancients.

The people of former eras had many of the problems which confront us today, and they sometimes solved them in almost the same manner. For instance, the ancient Romans would change some street arteries to one-way traffic during peak hours. The city of Pompeii used arm-waving traffic policemen to cope with the congestion. Street signs were used in Babylon more than 2,500 years ago, with curious names as, for example, the "Street on which may no Enemy ever tread." At Nineveh, the capital of Assyria, the following "no parking" signs were displayed: "Royal Road—let no man obstruct it." The signs were certainly more effective than ours, because instead of a parking offense ticket the chariot owner got a death warrant!

The ancient city of Antioch was the site of the first street

communication between the Mediterranean and the Red Sea was cut off until 1869.

Like the story of the Suez Canal the history of navigation has had a number of interesting pages. Modern Italian shipping companies must have got the idea of luxury liners from the ancient Romans. Two Roman ships found in the twenties at the bottom of Lake Nemi in Italy were restored between 1927 and 1932. The vessels were large and wide with four rows of oars. Accommodation was provided for one hundred and twenty passengers in thirty cabins with four berths in each, as well as quarters for the crew. The boats were richly decorated with mosaic floors depicting scenes from the *Iliad,* walls of cypress paneling, paintings in the lounge, and a library. A sundial in the ceiling showed the time, and it is thought that a small orchestra entertained the passengers in the salon.

The stern contained a large restaurant and a kitchen. The passengers enjoyed freshly baked bread for their breakfasts and the menus of the meals must have been comparable with the richness of the dining-room decoration. Certain finds came as a complete surprise—copper heaters provided hot water for the baths and the plumbing was absolutely modern, particularly the bronze pipes and taps. Centuries later Columbus or Magellan would not have dreamed of such ships!

The Roman patricians sailing on pleasure cruises in the Mediterranean certainly enjoyed the *dolce vita.* By a strange whim of fate these two Roman ships were destroyed—not by Carthage but by German bombers in the final stages of World War II. Evidently, their realistic contours tricked experienced pilots into believing that they were flying over barges under construction.

According to Chinese chronicles the Buddhist scholar Fa-hien returned from India around A.D. 400. He sailed from Ceylon directly to Java and then to northern China across the China Sea. The ship carried more than two hundred passengers and crew, and was larger than the vessels of Vasco da Gama crossing the Indian Ocean over 1,000 years later.

In a document called "Fusang," which was part of the annals of the Chinese Empire for A.D. 499, the Chinese Buddhist priest Hoei-shin related the story of his travels to distant lands. This country, where the monk landed after crossing the Pacific, is thought to have been Central America. As a matter of fact, in the last century a Chinese pirate junk reached California. It was displayed at Catalina Island near Los Angeles.

In 1815 a Japanese junk that had drifted in the Pacific Ocean for seventeen months was found near Santa Barbara, California. By a miracle, one sailor survived. After all, the story of Hoei-shin could be true.

The Great Wall of China is the longest wall that has ever been built on the face of the earth. It was constructed by three million workers in thirty-seven years about twenty-two centuries ago. The wall's length is 2,414 kilometers and it rises from 6 meters to 15 meters above the ground. The wall is wide enough to allow a lane of cars in each direction.

In 3100 B.C. King Menes of Egypt carried out a vast engineering scheme of diverting the course of the Nile in order to build his capital of Memphis. No nation had ever attempted to execute so gigantic a project as this.

Although porcelain flush toilets are not necessarily a mark of a high culture, they do prove the presence of a developed technology and sanitation. Only 200 years ago they were conspicuous by their complete absence. Yet 4,000 years ago private toilets with a central system of stone drains and ceramic pipes were common in the city of Knossos, Crete.

The rooms of the palace of Minos were ventilated through air shafts. With its air-conditioned chambers, excellent bathrooms, and toilets, the palace was not only "modern" but large—as large as Buckingham Palace.

Pipes for hot and cold water have been found in tiled

bathrooms at Chan Chan, the capital of the Chimu Empire in South America, which flourished in the eleventh to the fifteenth centuries. This technological achievement was nonexistent in Europe during the period of Richard Coeur de Lion or Jeanne d'Arc.

Ancient epics of India describe scientific accomplishments of the early people of the land of the Ganges. These tales cease to be legends once we realize the ingenuity of Oriental artificers.

The cave paintings of Ajanta near Bombay are admired by foreign tourists and Indian visitors alike. A great deal has been written about the excellence of these works of art but little has been said about the luminous paints of these murals. In one of the sixth-century catacombs there is a picture portraying a group of women carrying gifts. When the electric light is on, the beautiful painting lacks depth. But then the guide switches off the lights and the onlookers remain in darkness for a few minutes until their eyes become accustomed to it. Gradually, the figures on the wall appear to be three-dimensional as if they were made of marble. This fantastic effect was obtained by the ancient artist by the clever employment of luminous paints, the secret of which has been lost forever.

A number of soapstone columns stand in a twelfth-century temple in Halebid, Mysore. There are polished strips on one of these rough-finish stone pillars. When a person looks into the mirrorlike surface, he sees two reflections at the same time— himself in both an upright and an upside-down position. The unknown craftsman must have studied optics in order to have created so extraordinary an effect. In the city of Ahmadabad, Hujerat, there are two eleventh-century minarets in front of which stands an arch with a laconic inscription: "Swinging towers. Secret unknown." The height of the minarets is 23 meters and the distance between them 8 meters. When a group of visitors reaches the top of one tower, the guide climbs to the balcony of the other, grips the railing, and begins to swing his minaret. Immediately the other tower commences to sway to the amusement or alarm of the guests. These remarkable facts show that the roots of science are buried deep in time.

In the *House of the Four Styles* in the ruins of Pompeii an ivory statuette of the Indian goddess Lakshmi was discovered in 1938, which implied that commercial and cultural ties with India must have been maintained by Rome.

If, as the author has, you have traveled and seen the shops of Madras and Bombay, full of colorful saris, you may be surprised to find out that during the reigns of Vespasian and Diocletian textiles from India were on sale in Rome. But only the very rich could afford them. For silks, brocades, muslins, and cloth of gold bought in India, Rome remitted annually a considerable sum—possibly an equivalent of $40,000,000.

Silk, produced in China since the year 2640 B.C., was imported into ancient Rome in the first century A.D. Because of the long distance and the risks involved in transport, it was sold for an astronomical price in Rome.

One of the Seven Wonders of the Ancient World was the 135-meter-high Alexandrian Lighthouse on the island of Pharos, built of white marble. The tower had a movable mirror which at night projected its light so that it could be seen 400 kilometers away. Sunlight was used during the day, and fire during the night. The lighthouse stood from 250 B.C. until 1326, when an earthquake destroyed it.

These achievements of the people of antiquity were not surpassed in later centuries. In the Dark Ages mankind experienced a fall in scientific progress, and it is only during the last three hundred years that science began to pick up again.

No race had ever built 5,000-kilometer highways as the Peruvians did. They crossed canyons and pierced mountains with tunnels that are still in use today.

The first cart and the first boat were built by the Sumerians in the fourth millennium B.C. The next big leap in means of transport came only in 1802 when the steam vessel was constructed, and the first train followed in 1825. This acceleration in technology and transport was climaxed by the invention of the airplane in 1903 and the first manned flight in a spaceship in 1961.

After the voyage of Apollo 8 to the moon, the New York *Times* gave the real credit for this historic feat to "men of many countries and centuries—Euclid, Archimedes, Newton, Kepler, Copernicus, Tsiolkovsky, Oberth, Goddard and many others." It is wise to see our achievements in this light because behind our atomic scientists stands Democritus. Our aviation and astronautics engineers had a predecessor in antiquity— Heron with his jet. Back of our cyberneticians hovers Daedalus with his automatons and robots. The source of modern science lies far away in time.

3

Discoveries Create Problems

Paleontologists and archaeologists have produced a number of curious finds which still await a logical explanation. The story of man will appear in a different light if the answers are ever found. If the following facts are well founded, civilization might have had a much earlier source.

In excavations at Chou-kou-tien near Peking Dr. F. Weidenreich discovered a number of skulls and skeletons in 1933. One skull belonged to an old European, another to a young woman with a narrow head, typically Melanesian in character. A third skull was identified as belonging to a young woman with the distinctive traits of an Eskimo. A male European, a girl from the tropics, and another from the Arctic Circle uncovered on a Chinese hillside! But how, in the first place, did they get to China some 30,000 years ago? This episode out of prehistory is a mystery.

Did man in the last Ice Age possess enough technical facilities to straighten out a giant hooked mammoth tusk? Until the recent discovery of spears made of mammoth tusks by Dr. O. N. Bader near Vladimir, USSR, no scientist suspected that prehistoric man had possessed the ability to transform a hooked tusk into a number of straight bone spears.

On the same site the Russian archaeologists found a bone needle—a replica of our own steel needle. Like the spears it was 27,000 years old. The fact of the making of such artifacts by the Ice Age man was completely unexpected, and it entailed a reevaluation of views on technology in the Glacial Age.

The famous Jericho skulls, filled in with clay and shell, depict exquisite Egyptian-like faces. They have been dated to about 6500 B.C., which is roughly some 1,500 years before the beginning of Egyptian civilization. This discovery poses many questions. Were their mummified faces the outcome of a desire to immortalize man? If so, it provides evidence of the existence of religion in a very early period. But abstract thinking does not come to man overnight—it is a long process. From what source did the Jericho people receive it?

Professor Luther S. Cressman of the University of Oregon came across two hundred pairs of woven fiber sandals in Lamos Cave in east Nevada. Skillfully made by an artisan, they might be taken for modern beach sandals worn in St. Tropez or Miami. When a carbon-14 test was made, their age was shown to be well over 9,000 years.

But these sandals are young indeed when compared with the shoeprint discovered in a coal seam in the Fisher Canyon, Pershing County, also in Nevada. The imprint of the sole is so clear that traces of a strong thread are visible. The age of this footprint was estimated to be over 15,000,000 years. But man did not appear for another 13,000,000 years. Or in other words, according to current opinion, primitive man appeared some 2,000,000 years ago, but he only began to wear shoes 25,000 years ago! Whose footstep can it be?

Dr. Chow Ming Chen made a similar discovery in the Gobi

Desert in 1959. It was a perfect impression of a ribbed sole on sandstone and was calculated to be millions of years old. The expedition could not explain it.

The Brandberg rock paintings in South-West Africa depict Bushmen together with white women. Their perfectly European profiles are painted with a light tint, and the hair is shown in red or yellow. The girls wear jewelry and an elaborate headdress of shells or stones. The attractive young huntresses carry bows and waterbags on their chests. They are wearing shoes while the Negroes are not. Some archaeologists consider these young women to be brave travelers who must have come from Crete or Egypt 3,500 years ago. However, there is something peculiar about these white girls. They look like Caspians from north Africa who lived over 12,000 years ago. Both have the same long torsos, bows, headdress, and garterlike crossbands on their legs.

The White Lady of Brandberg studied by the Abbé Henri Breuil is a masterpiece. Because of her costume and a flower in her hand, she resembles a girl bullfighter of Crete. But for some reason no leopards or hippopotamuses are painted in this art gallery. These beasts were nonexistent in that part of Africa a long time ago, whereas they are quite common now. This circumstance opens a possibility that the epoch of the white Amazons in Africa may be more remote.

On a rocky cliff west of Alice Springs, in the heart of Australia, Michael Terry discovered a carving of the extinct *Nototherium mitchelli* in 1962. This rhinoceros-type animal had disappeared some 2,500 years ago. In the same place he also found six carvings of what looked like rams' heads. They brought to his mind Assyrian pictures of the ram.

A human being about two meters tall was among the intriguing rock images. The full legs and thighs, and a miter, resembling those worn by pharaohs, made the figure totally unlike the match-stick representations of human form drawn by the Australian aborigines. Though the figure is in a horizontal position, it is standing as if walking down a wall.

So here we have another mystery—carvings of the extinct Australian rhinoceros, the ram, unknown in Australia until the arrival of the English, and a non-Australian man in a Babylonian or Egyptian tiara. Signs of erosion of the rock carvings speak for their great age. Did men from the Near East or Asia reach Central Australia in antiquity, and if so, by what means? It seems that our views on the extent of the travels of ancient man should be amended.

As man is a recent evolutionary development (approximately 2,000,000 years old), his coexistence with monsters which lived thousands or even millions of years ago is considered by sicence to be impossible.

However, Professor Denis Saurat of France has identified the heads of animals in the calendar of Tiahuanaco in South America as those of toxodons, prehistoric animals which lived in the Tertiary period, many millions of years ago.* According to American writer and archaeologist A. Hyatt Verrill, the Cocle ceramics of Panama depict a flying lizard which looks very much like the pterodactyl that lived eons before man.†

In 1924 the Doheny Scientific Expedition discovered in Hava Supai canyon, in northern Arizona, a rock carving which looked amazingly like the extinct tyrannosaurus standing on its hind legs. In another rock image in Big Sandy River in Oregon, the prehistoric sculptor left a portrait of a stegosaurus, a creature which lived before the appearance of man on this planet.

The drawings, made by the scratching of red sandstone with a flint, show signs of great age. The existence of the artists must have been contemporaneous with that of the prehistoric monsters, otherwise how could primitive man draw beasts

*D. Saurat, *Atlantis and the Giants* (Faber & Faber, London).

†A. H. Verrill, *Old Civilizations of the New World* (New Home Library, New York, 1943).

which he had never seen? Naturally, these impossible facts menace the whole structure of anthropology.

About 1920 Professor Julio Tello found vases in the Nasca district near Pisco, Peru. He was struck by the figures of llamas painted on the vessels as the animals are shown with five toes. At the present time, the llama has only two toes but in an early evolutionary stage, tens of thousands of years ago, it did have five. This is not a mere conjecture because skeletons of the prehistoric llama with five toes have been excavated in the same region.

The discovery of megalithic sculptures in Marcahuasi by Dr. Daniel Ruzo in 1952 was a momentous one. Marcahuasi, about 80 kilometers northeast of Lima, Peru, is a plateau at an altitude of 4,000 meters, where the air is cold and hardly anything grows amidst the granite rocks.

Standing in an amphitheater of rock, Ruzo found himself confronted by the enormous figures of people and animals carved out of stone. Caucasian, Negro, and Semitic faces looked at him. Lions, cows, elephants, and camels, which had never lived in the Americas, surrounded Dr. Ruzo. He spotted the amphichelydia, an extinct ancestor of the turtle known only through its fossilized remains. Sculptures of the horse posed a burning question—was the carver a contemporary of the American horse? Since the horse died out in the Americas about 9,000 years ago, this gave a definite date to these ancient works of sculpture. The horse reappeared in the New World only in the sixteenth century, when the conquistadors brought it from Spain.

By analyzing the white dioritic porphyry from which the heads were carved, geologists arrived at the conclusion that the stone would have needed at least 10,000 years to take on the gray tint it now shows in the cuts.

The mysterious sculptors of these giant monuments were aware of the laws of perspective and optics. Some figures can be seen at noon, others at other times, vanishing as the shadows move. To find a 10,000-year-old museum exhibiting animals which either never lived in South America or had been extinct for tens of thousands of years, as well as sculptured portraits of the Negro and the white man, who came to the New World within the last 500 years, was a challenge to orthodox science. Dr. Ruzo has lectured at the Sorbonne and at other scientific institutions. Although official circles, having seen the photographs of these sculptures, could hardly deny the fact of this amazing discovery, they questioned Ruzo's theory that races other than the Red Indian had lived in South America. However, whoever the rock sculptors were, their coexistence with extinct animals cannot be doubted.

A strange discovery was made in Costa Rica, also in the fifties. Hundreds of perfectly shaped spheres made of volcanic rock were scattered in the jungle. Their sizes ranged from 2½ meters to a few centimeters. A number of the larger ones weighed as much as 16 tons. Similar globes are also located in Guatemala and Mexico but nowhere else in the world.* These balls have raised many questions. What ancient race could have carved and polished them so perfectly? The technical difficulties in making these spheres and transporting them to the sites would have been enormous. What was the purpose of the stone balls? Or are they natural geological formations, as some scientists believe? Some of the balls rest on stone platforms, which seems to indicate that they were placed there for some reason. Many globes are arranged in clusters, in straight lines, or in a north-south direction. There is an indication of a geometrical pattern because some groups form triangles, squares, or circles. It has been suggested that these megalithic markers might have some astronomical significance. It would be interesting to draw a complete map showing the location of these globes and then to see whether there is any resemblance to the constellations on a star chart.

Natural History (U.S.A.), September, 1955.

However there is the alternative theory that the stone balls were used for astronomical observation in the manner of the Stonehenge megaliths.

The giant stone heads of the Olmechs found in La Venta, Tres Zapotes, and other sites in Mexico can be classed as artifacts of a similar type. These colossal heads carved of black basalt are from 1.5 to 3 meters high, weighing from 5 to 40 tons. They are placed on stone stands just like the globes described above. The nearest basalt quarries are 50 to 100 kilometers away. How could a people without wheeled vehicles or pack animals bring these masses of rock across swamps and jungles to the erection sites? These immense faces of La Venta and San Lorenzo have been dated 1200 B.C.—another surprise for the historians of science.

But let us put these stone heads aside and speak of real skulls. On the ground floor of the Museum of Natural History in London, a human skull is displayed. It comes from a cavern in Northern Rhodesia and has a perfectly round hole on the left side. There are no radial cracks, which are usually present if an injury is caused by a cold weapon. The right side of the skull is shattered. The skulls of soldiers killed by rifle bullets have an identical appearance. The cranium belongs to a man who lived over 40,000 years ago, at a time when no guns were made. An arrow could not have produced such a perfectly round hole on the left side of the skull and shattered the right side as well.

The Paleontological Museum of the USSR has a skull of an auroch which is hundreds of thousands of years old. It shows a clear round hole on its frontal part and scientific evidence has proved that although the skull was pierced, the brain was not injured and the beast's wound healed. In that distant past, the ape-man was supposedly armed only with clubs. The perfectly round hole without radial lines looks very much like one made by a bullet. The question is—who shot the auroch?

A meteorite of an unusual shape found near Eaton,

Colorado, created a riddle. An analysis by an American expert on meteorics, H. H. Nininger, indicated that the meteorite was composed of an alloy of copper, lead, and zinc (that is, brass) which does not exist in nature. The meteorite could not have been "space garbage" because it fell in 1931.

In the sixteenth century the Spanish conquistadors came across an 18-centimeter iron nail solidly incrusted in rock in a Peruvian mine. The rock was estimated to be tens of thousands of years old. Since iron was unknown to the American Indian until the Conquest, one wonders whose nail it was. The Spanish Viceroy Don Francisco de Toledo kept the mysterious nail in his study as a souvenir.*

According to the London *Times* of December 24, 1851, Mr. Hiram de Witt found a piece of auriferous quartz in California. When he dropped it accidentally, an iron nail with a perfect head was found to be inside the quartz. About the same time Sir David Brewster made a report to the British Association for the Advancement of Science which created a sensation. A block of stone from Kingoodie Quarry in north Britain contained a nail, the end of which was corroded. But at least an inch of it, including the head, lay embedded in the rock. Because of the great age of the geological strata where these three iron nails were found, the identity of their makers remains a mystery.

In 1885 in the foundry of Isidor Braun of Vöcklabruck, Austria, a block of coal was broken and a small steel cube, 67 mm by 47 mm, fell out. A deep incision ran around it and the edges were rounded on two faces. Only human hands could have made these. The son of Braun took the article to the Linz Museum but in the course of decades it was lost. However, a cast of the cube has been kept by the Linz Museum. Contemporary magazines, such as *Nature* (London, November, 1886) or *L'Astronomie* (Paris, 1887), had articles about this strange find. Some scientists endeavored to explain

*Madrid Archives, letter of October 9, 1572.

it as a meteorite from the Tertiary coal period. Others wanted an explanation for the groove around the cube, its perfect form, and the rounded edges, and claimed that it had an artificial origin. The debate has never been closed.

These perplexities cannot be cleared up unless a reappraisal of prehistory is made. The facts assembled here point to the existence of a technology at what we have imagined to be the dawn of mankind. Two theories can explain the artifacts described in this chapter—either there was some kind of technological civilization in a bygone past, or the earth has been visited by beings from other stellar worlds.

The true significance of many museum exhibits may have evaded our comprehension. These cryptograms in marble, stone, wood, or bronze may carry a significant message. In 1946 the Carnegie Institution reported an archaeological find in Kaminaljuyu, Guatemala—a peculiar 32-centimeter figurine of a mushroom with a human face, with widely opened eyes, at the root. The meaning of the object was obscure. But when Spanish records of the "sacred mushrooms" and their use by the Mexican priests had been studied, experimenters decided to try these mushrooms. A state of narcotic trance with psychedelic visions was produced. The figurine gives the whole story symbolically.

The birth of metallurgy, chemistry, medicine, physics, astronomy, technology, and other wonderful accomplishments of the ancients will be outlined in the following chapters.

4

The Blacksmith of Olympus

Technology began with Hephaestus, or Vulcan, the world's first metallurgist, according to Greek mythology. His workshop—a sparkling dwelling of bronze—was on Mount Olympus. But eventually he settled in Sicily on Mount Etna, and legends affirm that the smoke from the crater comes from the furnaces of the god. Although the author has seen this smoke from Taormina, he could not find out whether Hephaestus was still at his anvil.

Greek myths speak of the four ages of man—first came the Golden Age, followed by the Silver Age, after which arrived the Bronze Age. The last epoch is the Iron Age in which we live today.

Although iron is more plentiful than copper or gold, it is more difficult to melt and forge. Thus the ancient Greeks told us about the progress of metallurgy by this simple tale of how it had started with soft metals and ended with hard iron.

The Stone Age, which had lasted for a long time, was

followed by the Chalcolithic Age, when the old, perfected stone implements were mainly used but copper tools and weapons were also making their appearance as luxuries.

Then came bronze, a hard alloy made of copper with the addition of one-tenth part of tin. The third millennium B.C. in Sumeria and Egypt is predominantly the Copper and Bronze Age. No clear picture is available of where and how the bronze first appeared. To combine copper, which came from Sinai, Crete, Cyprus, Spain, Portugal, or other parts of the Mediterranean, with rare tin from Etruria, Gaul, Spain, Cornwall, and Bohemia, it would have been necessary to have organized transport, skilled labor, and furnaces with temperatures well over 1,000° C.

Bronze, a mixture of copper and tin, is strong and durable. It should have taken long ages to discover that the addition of one-tenth part of tin to copper creates a better metal. Yet strangely enough, copper artifacts in our museums are few. Bronze seems to have appeared suddenly and spread far and wide in great profusion.

The similarity of bronze articles found in different parts of Europe compels us to conclude that they came from one manufacturing center or school of technology.

The history of bronze in Central and South America is similar. The alloy appears quite suddenly. Was the discovery made by experimentation or by chance?

The discovery of bronze was not simultaneous in the Old World and the New. Copper, which is a component of bronze, was mined in Mesopotamia about 3500 B.C. but not before 2000 B.C. in Peru. (Iron was unknown to the Incas until after the arrival of Pizarro.)

Certain achievements of the South American Indian in metallurgy are enigmatic. Ornaments of platinum were found in Ecuador. This poses a provoking question—how could the American Indian produce the temperature of over 1,770° C necessary to melt it? It should be borne in mind here that the melting of platinum in Europe was achieved only two centuries ago.

In testing an alloy from a prehistoric artifact the United States Bureau of Standards ascertained that the original dwellers of America had furnaces capable of producing a temperature of 9,000° C 7,000 years ago. No satisfactory explanation has yet been given of how such a technical feat was possible at all at so remote a date as 5000 B.C.*

The tomb of the Chinese general Chow Chu (A.D. 265-316) presents a mystery. When analyzed by the spectroscope, a metal girdle showed 10 percent copper, 5 percent manganese, and 85 percent aluminum. But according to the history of science aluminum was obtained for the first time by Oersted in 1825 by a chemical method. To satisfy industrial demands, electrolysis was later introduced into the manufacturing process. Needless to say, an ornament made of aluminum, whether chemically or electrolytically produced, seems out of place in a third-century grave in China. It is hardly reasonable to think that this aluminum article was the only one manufactured in China.

The Kutb Minar iron pillar in Delhi weighs 6 tons and is about 7.5 meters high. For fifteen centuries it has withstood the tropical sunshine of India plus the heavy downpours during the monsoons. It does not show any signs of rust formation, and provides proof of the superior metallurgical skill of ancient India. Aside from the mystery of the noncorrosive metal of which the column is made, the task of forging so large a pillar could not have been achieved anywhere in the world until recent times. The production of rustproof iron of this type is possible today because of our high technology but it is surprising to find such an achievement in A.D. 415. The pillar stands as a mute witness to the scientific

*Science et Vie, No. 516.

50

tradition preserved by the people of antiquity in all parts of the world.

Men whom time has forgotten held the answers to these riddles of the history of science.

5

The Forgotten Art of Goldmaking

Alchemy was modern chemistry in ancient garb. But it was also the art of transmutation of base metals into precious ones.

For many centuries scholars thought that chemical elements were stable and could not be transformed. This is why the alchemists were regarded as dreamers, charlatans, or idiots. But in the year 1919 the great English physicist Rutherford sided with the alchemists and transmuted nitrogen into oxygen and hydrogen by bombarding it with helium. That was the day of the vindication of the alchemical doctrine of transmutation.

Alchemy, as a controlled transformation of one element into another, was the subject of prolonged study by the Orient as well as the Occident, gradually giving birth to modern chemistry. There are extant medieval manuscripts which describe in detail the equipment of the alchemists, comprised of retorts, glass vessels, distilling stills, furnaces, and other things necessary for the "Great Work." The cost of an average chemical laboratory must have been considerable.

It is absurd to suppose that all these goldmakers parted with their coin to sweat for months and years near their furnaces without a hope of getting some tangible results from their work. Although there were individuals who abandoned alchemy after having failed to transmute cheap metals into gold, the number of people who persevered in this art throughout their lives was surprisingly great.

In view of the costly laboratory equipment and materials required for transmutation work, how could they afford it without reaping a profit of some kind?

Down through the centuries alchemists have claimed they could perform transmutations of mercury, tin, or lead into gold. Those who believe that anything the ancients could do, we can do better, will naturally express doubt as to the ability of the alchemists to accomplish this scientific feat. Wasn't alchemy a charlatanry of some sort? Actually, history does mention the names of men who tried to capitalize on the credulity and greed of their contemporaries. On the other hand, there are historical documents dating back many centuries which demonstrate that rulers often considered alchemists to be a menace to the economy of the state.

The Roman emperor Diocletian issued an edict in Egypt around A.D. 300, demanding that all books on "the art of making gold and silver" be burned. The decree shows that the Roman government was certain that such an art of transmutation of metals had existed. It would surely have been unnecessary to issue decrees banning this craft unless it were known to have been practiced.

This same emperor signed an order to destroy all secret and open places of Christian worship as well as Christian book. All the Christians were removed from official posts in the Roman Empire. Rome meant exactly what it had stated in the government proclamation.

The decree against alchemy and its practitioners was of the

same type, and presumably the existence of artificially produced gold was taken for granted as was the presence of Christians. The Roman emperor wanted to withdraw all written records of this secret art from circulation. It is not difffcult to ascertain the motives of Diocletian. He realized that gold was power. An alchemist capable of making it cheaply could become a threat to the state. Such a man could buy territories or officials. It is worthwhile citing the earlier case of the Praetorian Guard Didius Marcus, a Roman millionaire, who bought the whole Roman Empire for the equivalent of about $35,000,000. However, he was soon beheaded by Emperor Septimus. This historical episode was still fresh in the minds of the Roman citizens when Diocletian issued the prohibition against alchemy.

According to the alchemist Zosimus (A.D. 300), the temple of Ptah at Memphis had furnaces, and this god was revered as the patron of the alchemists. The words *chemistry* and *alchemy* are derived from the name of Egypt—*Khemt.* Thus even today a very ancient tradition is perpetuated by the use of these words—*alchemy, chemistry, chemist,* or *chemical.*

In the eighth century the Arab Jabir (Geber) systematized alchemical knowledge from the Egyptian source, and he is justly called the father of this science. Jabir was a practicing alchemist who described not only the equipment of a laboratory required for transmutation but also the mental and moral prerequisites of an apprentice. "The artificer of this work ought to be well skilled and perfected in the sciences of natural philosophy," wrote the Arab scholar. Considering the time and labors involved in discovering the secret of transmutation, Jabir advised the disciple not to be extravagant, "lest he happen not to find the art, and be left in misery."

It goes without saying that the Arab adept spoke of very concrete things—a chemical laboratory and patient efforts which would not pay dividends for years to come. But he assured the students that "copper may be changed into gold" and "by our artifice we easily make silver." These statements cannot be easily dismissed as Jabir's name figures in the history of modern chemistry.

One of the peculiarities of alchemy was its extensiveness. Alchemy was known in China as early as 133 B.C. The story of Chia and the alchemist Chen mentions that whenever Chia wanted money, his friend the alchemist would rub a black stone on a tile or a brick and transform these commonplace articles into precious silver. That was an easy way to make money.

The biography of Chang Tao-Ling, who studied at the Imperial Academy in Peking, makes reference to the *Treatise of the Elixir Refined in Nine Cauldrons,* which he found in a cavern and whose author was allegedly the Yellow Emperor (twenty-sixth century B.C.).

The basic ingredient of Chinese alchemy was cinnabar or mercuric sulfide, used in transmutation as well as in the preparation of "gold-juice," the elixir of youth. "You may transmute cinnabar into pure gold," assures the historical record *Shih Chi,* written in the first century B.C.

The opinion current among the practitioners of the alchemical art in China, India, Egypt, and Western Europe— that mercury and sulfur had unusual properties for transmutation—is really baffling. After all, it was a long way from Peking to Alexandria, and from Benares to medieval Paris. What was the primary source of this doctrine?

A law was enacted in China against the practice of counterfeiting gold by alchemical methods in 175 B.C. This fact proves two things—firstly, alchemy must have existed in China for many centuries before becoming a problem to the Celestial Empire, and secondly, the output of gold by the alchemists was sufficiently large to be felt by the state.

India had alchemy, too. The Hindu expositors of the art also thought that mercury and sulfur were primary elements.

55

But unlike Chinese and European alchemists they attributed positive polarity to mercury and negative to sulfur. They also tried to discover the elixir of immortality and the secret of goldmaking.

In view of the fact that the art of transmutation and the production of gold placed its adepts in a dangerous position because of envy, malice, possible robbery, and even loss of life, to say nothing of the suspicion of the authorities, the alchemists used carefully coded texts and enigmatic charts. This is particularly true of European countries where the Inquisition was busy tracking down and liquidating anyone guilty of practicing the "magical sciences" from the heathen East.

The question as to whether gold had been produced by alchemical processes in the past can be hotly debated. But certain decrees and documents imply that the rulers of many nations did not have any doubts about the possibility of the transmutation of metals. This is good evidence of the reality of alchemy in olden times.

During the thirteenth and early fourteenth centuries alchemy must have been widespread, as it attracted the attention of the Vatican. The science was forbidden by a bull of Pope John XXII in the year 1317. This document, entitled *Spondent Pariter,* condemned the alchemists to exile and established heavy fines against swindlers commercializing transmutation.

All these prohibitions of alchemy are very bewildering. A "No Smoking" sign in a train is put up because people have cigarettes in their pockets. What was the reason for these "No Goldmaking" orders? If there were no cases of illegal transmutation, it surely would not have been worthwhile wasting expensive parchment on long, sternly worded decrees.

Henry IV of England issued an act in 1404 declaring that the *multiplying of metals* was a crime against the Crown. This was during the time of the Hundred Years' War and the

Peasants' Revolt. A King of England was not likely to sign a decree against a mythical menace while waging a very real war in France and fighting angry serfs at home. Apparently, the appearance of gold from an unknown source began to worry the English government.

On the other hand, King Henry VI granted permits to John Cobbe and John Mistelden to practice "the philosophic art of the conversion of metals," and these licenses were duly approved by Parliament. This alchemically made gold was used in coinage, which makes it clear that the Crown did not mind the manufacture of alchemical gold provided the Royal Mint received it in the end.

But much more significant than Henry IV's ban on alchemy was its official repeal by William and Mary of England in 1688, which reads: "And whereas, since the making of the said statute, divers persons have by their study, industry and learning, arrived to great skill and perfection in the art of melting and refining of metals, and otherwise improving and multiplying them."

The act of repeal states that from the reign of Henry IV many Englishmen went to foreign countries "to exercise the said art" to the great detriment of the kingdom. The new decree announced that "all the gold and silver that shall be extracted by the aforesaid art" be turned over to the Royal Mint in the Tower of London, where the precious metals would be bought at the full market value, and no questions asked.

After this change of policy the king and queen even made a declaratcion concerning the desirability of studying alchemy. These historic facts are most extraordinary because alchemically made gold might be stacked in ingots in the vaults of the Bank of England today! It is important to note that, as far as we know, England has always received its gold supplies from foreign countries only. It is apparent that the sovereigns of England realized that there were advantages in controlling gold reserves rather than permitting this gold from an unknown source to dominate the economy of the realm. This repeal act clearly states that artificially manufactured gold was actually produced in England and also that its intake was centralized at the Royal Mint.

This possibility of artificial gold's having been produced in England is well substantiated by a specimen of alchemical gold which the author has personally examined in the Department of Coins and Medals of the British Museum in London. It is in the form of a bullet, which is understandable as that is what it was before the transmutation. The register of the museum contains the following brief entry concerning this golden bullet: "Gold made by an alchemist from a leaden bullet in the presence of Colonel MacDonald and Doctor Colquhoun at Bupora in the month of October, 1814."

Although the information about the actual transmutation is lacking, the fact remains that this is officially recognized as a rare specimen of alchemical gold, preserved in one of the world's greatest museums.

Johann Helvetius (1625-1709), physician to the Prince of Orange, was known to have accomplished alchemical transmutations of base metals into gold. Once Porelius, the inspector-general of the mint in Holland, came to Helvetius' laboratory to watch his alchemical work. Then Porelius went to see the jeweler Brechtel and asked him to make an analysis of Helvetius' gold. After a rigid test the gold was found to have more grains than before the test.

Now what is transmutation? Plutonium, an element which is nonexistent on earth, can be created by nuclear physics— that is a case of transmutation. A hypothetical transmutation of mercury into gold would involve changing the atomic structure of mercury. The number of electrons, their orbits, and the organization of protons determines the element. It is noteworthy that, according to ancient alchemy, gold was made from mercury or lead. In the periodic table of elements the atomic number of gold is 79, that of mercury 80, and of lead 82—in other words, they are neighbors. It was Mendeleyeff who in 1879 first formulated a table of the elements and arranged them in order of increasing weight according to their atomic structure. The question is—had the alchemists discovered this table before Mendeleyeff?

Arab scholars such as Jabir, Al Razi, Farabi, and Avicenna, who lived between the eighth and the eleventh centuries, brought the science of alchemy to Western Europe. Costly handwritten books were carried from city to city. They contained ciphered writings and mysterious diagrams which few could read and fewer understand. Some of these manuscripts and tracts embodied true chemistry and alchemy, others but distorted versions of ancient formulas and methods of no practical value.

The alchemists drifted from place to place, practicing their art in secret. It was dangerous to declare one's proficiency in transmuting cheaper metals into gold because sovereigns often subjected talkative men to torture in order to obtain the alchemists' formulas. In the *Compound of Alchemy* (1471) Sir George Ripley advised the students and practitioners of the art "to keep thy secrets in store unto thyself for wise men say store is no sore."

The pioneers of modern science such as Albertus Magnus (1206-1280), who wrote voluminously on astronomy and chemistry, not only believed in the reality of alchemical transmutation but even made rules on how to practice the art. His advice was "to carefully avoid association with princes and nobles and to cultivate discretion and silence."

Roger Bacon (c. 1214-1294) left a ciphered manuscript which Professor W. R. Newbold has allegedly decoded. It contains a formula for making copper. In the library of the University of Pennsylvania there is a retort and the following certificate, dated December 1, 1926: "This retort contains metallic copper made according to a secret formula of Roger Bacon."

The great doctor Paracelsus (1493-1541) discovered zinc and was the first to identify hydrogen. Paracelsus' fame as an alchemist was so great that his tomb in Salzburg was opened because of rumors that alchemical secrets and great treasures had been buried with the physician. However, nothing was found in the coffin. His famous sword, whose hilt contained the so-called philosopher's stone, had also vanished without a trace.

Nicolas Flamel (1330-1418), a Paris notary, was another great alchemist. In his business of illuminating documents and manuscripts he came into contact with bookdealers. In his *Hieroglyphical Figures* he related that a very ancient book of Abraham the Jew, written in an unknown language, was offered to him for sale by a stranger for a reasonable amount, and that he bought it. It took Flamel and his wife, Pernelle, many years to come to the conclusion that the book was a work on ancient alchemy.

Using the text Nicolas Flamel was able to perform his first transmutation of one-half pound of mercury into pure silver on January 17, 1382, when he was fifty-two years old. On April 25 he succeeded in making his first alchemical gold.

The citizens of fourteenth-century Paris were less skeptical about Flamel's ability to manufacture gold than the Parisians of today. But they had good reason—the alchemist built many hospitals and churches in Paris during the thirty-six years of his profitable alchemical work. This fact he admitted himself: "In the year 1413 after the transition of my faithful companion whom I will miss for the rest of my life, she and I had already founded and endowed fourteen hospitals in this city of Paris besides three completely new chapels, decorated with handsome gifts and having good incomes, seven churches with numerous repairs done to their cemeteries, as well what we ourselves had done in Boulogne* which is hardly less than what we did here."

Nicolas Flamel wrote that on some of his churches he "caused to be depicted marks or signs from the Book of Abraham the Jew." They could actually be seen two hundred years ago in such places as the Cimetière des Innocents, the church of St. Jacques de la Boucheries and St. Nicolas des Champs. The Musée Cluny contains Flamel's tombstone.

The book of Abraham the Jew is probably not fictitious as it was listed in the *Catalogus librorum philosophicorum hermeticorum* issued by Dr. Pierre Borelli in 1654. Borelli was obviously no ordinary savant as he was farsighted enough to imagine "aerial ships" as the means "whereby one can learn the pure truth concerning the plurality of worlds."

According to Dr. Borelli, Cardinal Richelieu ordered a search for alchemical books in Flamel's house and churches which must have been successful because at one time the cardinal was seen reading the book of Abraham the Jew with annotations by Flamel in the margins.

The case of George Ripley, an English alchemist of the fifteenth century, was equally spectacular. Elias Ashmole, an English scholar of the seventeenth century who left a collection at Oxford known as the Ashmolean Museum, mentioned a document in Malta citing a record of contributions of £100,000 each year made by Sir George Ripley to the Order of St. John of Jerusalem at Rhodes to help Rhodes fight the Turks. It should be stressed that the value of the pound was immensely higher five hundred years ago than it is today.

Other alchemists were evidently making so much gold that one of them offered to finance a Crusade, and another to pay off the national debt of his country. With the monetary crises of today and deficits piling up yearly, finance ministers might

*Boulogne-sur-Seine.

do well to try calling alchemy to the rescue to build up gold reserves.

Pope John XXII, who issued a bull against the alchemists, developed an interest in the art himself! It is quite possible that after having perused numerous confiscated documents on alchemy, he decided to experiment in the science of transmutation. In fact, he wrote an alchemical work, *Ars Transmutatoria,* in which he related how he had worked on the philosopher's stone in Avignon, and how he had alchemically manufactured two hundred bars of gold, each weighing one quintal, or one hundred kilograms. After his death in 1334, twenty-five million florins* were found in the pope's treasure vault! The source of this vast fortune could never be satisfactorily explained, because in this era of wars and the ecclesiastical conflict between Avignon and the Vatican, the papal revenues were small.

The Kunsthistorisches Museum in Vienna contains extraordinary evidence of the practice of alchemy in past centuries. It is catalogued as an *Alchimistisches Medaillon—*an oval medal 40 by 37 centimeters in size weighing 7 kilograms. Except for the upper third, which is silver, the disk is solid gold.

This medal has an exciting story to tell. In an Augustine monastery in Austria there was a young monk in the sixteenth century whose name was Wenzel Seiler. He was bored with life in the abbey but without riches there was no way of his getting out. An old friar who patronized Wenzel had told him of a treasure buried in the monastery, so they decided to look for it.

After a long search they found an old copper chest under a column. It contained a parchment with strange signs and letters, and four jars of reddish powder. Seiler expected to find gold coins in the box and was so disappointed that he thought

*The fourteenth-century gold florin may be worth as much as $5 today.

of throwing out the contents. But the old monk became interested in the document and insisted that the powder be preserved.

The aged friar finally came to the conclusion that the red powder could be the precious transmuting compound of the alchemists. Then Wenzel Seiler stole an old tin plate from the abbey's kitchen and after covering it with the red powder he heated the plate in the fire. As if by magic, the tin plate shortly became solid gold!

Seiler was so happy with the results of the experiment that he went to town to sell the gold. He received twenty ducats for it but the old friar did not think it was a wise thing for a young monk to sell gold. The old man became sick and died soon after, leaving young friar Wenzel the sole possessor of the goldmaking powder.

Realizing that he was unable to exploit his discovery and escape from the monastery without assistance, he confided his secret to Francis Preyhausen, another young monk, and they made plans to leave the abbey in the spring.

With his ducats Wenzel bought wine and enjoyed the visits of his young cousin Anastasio from Vienna. Rumors about the stolen plate, the twenty ducats obtained from a jeweler, and the empty wine bottles reached the abbot, who summoned Seiler for questioning. Then the abbot with the older friars went to Wenzel Seiler's cell. They unlocked the door and saw naked Anastasio on Wenzel's bed. When they saw his anatomy, it suddenly dawned upon the aged monks that Anastasio was Anastasia! After a few embarrassing moments during which the girl had time to wrap herself in a cape, the men of God gave her a sermon on the dangers facing her soul.

But young Wenzel was flogged and bolted in his cell. The four precious jars with the red powder were surreptitiously handed through the bars of the window to Francis, who was waiting outside. Then Wenzel Seiler was transferred to a prison cell and the future began to look very dark. However,

Francis Preyhausen was not idle and he arranged their escape.

During an adventurous journey the young monks understood how dangerous their lives could become with the goldmaking powder in their hands. But Francis was more intelligent than Wenzel and he hid the powder.

In Vienna they secured the patronage of Count Peter von Paar, a friend of Emperor Leopold I of Germany, Hungary, and Bohemia (1640-1705), as the noble was an ardent student of alchemy. An audience was arranged with the emperor, who was also interested in the ancient art.

In the presence of Leopold I, Father Spies, and Dr. Joachim Becher, ex-friar Wenzel Seiler transmuted an ounce of tin into pure gold in the course of a quarter of an hour. A written declaration to that effect was signed by the witnesses.

Count von Paar's friendship was not as sincere as it had first seemed. With pistol in hand he forced Wenzel to part with a portion of the red powder. Fortunately for Wenzel and Francis, the nobleman died soon after the incident.

Emperor Leopold I then became Seiler's protector. With Count von Waldstein, the captain of the bodyguard, the emperor himself made alchemical gold with Wenzel Seiler's red powder. In 1675 a special ducat was struck with the image of Leopold I from the gold alchemically produced by the sovereign. On the reverse side was the following inscription:

> With Wenzel Seiler's powder
> Was I transformed from tin into gold.

Successful experiments in alchemy were conducted by Seiler at the Palace of the Knights of St. John in the Kärntnerstrasse in Vienna, and a gold chain was made from this alchemical gold on the orders of Count von Waldstein.

On September 16, 1676, the emperor knighted the alchemist-monk von Rheinburg, which was the maiden name of Seiler's aristocratic mother (as his father was a commoner), and appointed him Court Chemist.

With the red tincture almost gone, Wenzel Seiler and Leopold I concentrated their efforts on multiplying the powder but without any results. In 1677 a large silver medal was dipped into the transmuting compound and its lower part turned into gold. A photograph of the medal is featured in this book, and the only remark that has to be made about it concerns the four notches on its edge. These were made on the request of Professor A. Bauer of Vienna in 1883 in order to analyze the content of the disk. Two-thirds was found to be solid gold, so there was no question of any gold plating. This case of alchemy is recorded in history and offers strong evidence in support of the reality of alchemical transmutation in former times.

There is a nineteenth-century painting by the Polish artist Matejko which portrays dramatically an actual alchemical transmutation by Michael Sendivogius in Cracow before King Sigismund III of Poland, early in the seventeenth century.

Alchemy was not confined to making gold alone, as some alchemists claimed they could produce gems. If so, they must have been the first synthetic-stone makers.

Modern science can transform a lump of anthracite into an expensive diamond but the process is costly. Dr. Willard Libby, Nobel Prize winner, created diamonds by sandwiching a block of graphite between two nuclear devices in 1969. Dr. E. O. Lawrence of the United States effected transmutations of a number of elements during the forties.

In 1897 Dr. Stephen H. Emmens, a British physician in New York, claimed that he had discovered a method for transmuting silver into gold. Between April, 1897, and August, 1898, more than $10,000 worth of gold was sold by him to the U.S. Assay Office in Wall Street. The New York *Herald* printed the following headlines about Dr. Emmens at the time: "This Man Makes Gold and Sells It to the United States Mint." The Assay Office admitted buying the gold but at the same time raised the question: "Did he manufacture it out of silver as he claimed?"

It is of little consequence whether or not the alchemists could actually transmute silver, tin, or lead into gold. What is more significant is the fact of their thinking that one chemical element could be transformed into another. Until Curie and Rutherford science excluded this possibility. In brief, the alchemists anticipated our modern scientific concepts regarding the essence of matter.

In his *Interpretation of Radium* published in 1909, Dr. Frederick Soddy, Nobel Prize winner, who coined the word *isotope* and pioneered nuclear physics, did not deride alchemy:

> It is curious to reflect, for example, upon the remarkable legend of the Philosopher's Stone, one of the oldest and most universal beliefs, the origin of which, however far back we penetrate into the records of the past, we do not probably trace to its real source. The Philosopher's Stone was accredited the power not only of transmuting the metals but of acting as the elixir of life. Now, whatever the origin of this apparently meaningless jumble of ideas may have been, it is really a perfect and very slightly allegorical expression of the actual present views we hold today.

Egyptian tradition pointed to Thoth, Hermes, or Mercury, the culture bearer, who had revealed to mankind the Hermetic arts, one of which was alchemy. Hermes, or Mercury, was also the founder of medicine. It is upon the rock of Hermetic science that modern medicine is built. It is fascinating to trace the stream of medical science from prehistoric medicine man, herbalist, magician, priest to the pharmacist and doctor of contemporary life.

6

The Caduceus of Hermes

Doctors' cars usually carry an emblem—a staff with two snakes and a winged hat. This is the caduceus of Hermes and by this ancient symbol modern medicine acknowledges its debt to the sages of antiquity.

A recent archaeological expedition to the Valley of the Kings in Egypt excavated a number of mummies. Many of the jaws had bridges and artificial teeth which looked surprisingly like the product of a modern dentist. Few scientists had expected to find evidence of such skill in dentistry in ancient Egypt so many thousands of years ago.

Mayan skulls dug up on the coast of Jaina in Campeche, Mexico, also show astonishing proficiency in dental surgery. The crowns and fillings are still in place after many centuries! The drilling and setting of inlays was done by men who always respected the vital part of the tooth. The adhesives used are as yet unknown but they must have been of high quality if the fillings are still intact.

The pre-Inca surgeons performed delicate operations on the

brain 2,500 years ago. Trepanation is a new technique in modern surgery, so it was more than surprising to find thousands of skulls in Peru with marks of successful trephining. The instruments used were obsidian arrowheads, scalpels, bronze knives, pincers, and needles for sutures. According to the history of medicine, the same operation performed at the Hôtel Dieu in Paris in 1786 was invariably fatal.

Amputations were likewise executed in South America. The Inca doctors used gauze for dressings, and possibly cocaine as an anesthetic. The Incas discovered important drugs such as quinine, cocaine, and belladonna.

In ancient Babylon there was a peculiar method of treating the sick. Herodotus describes the way the sufferers were brought out into the street. It was the moral duty of passersby to inquire about their complaints. From their own experiences the sympathizers suggested remedies which they had heard were effective or had used themselves. By experimenting with different medicines the patients found out which were best for them. This mass experimentation formed the basis of pharmacopoeia and diagnosis in the centuries to follow.

Our wonder drugs like penicillin, aureomycin, or terramycin had their origin in ancient Egypt.* A medical papyrus of the eleventh dynasty speaks of a certain type of fungus growing on still water which is prescribed for the treatment of wounds and open sores. Did they have penicillin 4,000 years before Fleming?

Antibiotics were not unknown to the ancients. Warm soil and soybean curd, which have antibiotic properties, were employed by the ancient Greeks and the Chinese respectively—to heal wounds and to eradicate boils and even carbuncles.

The Egyptians made use of an unknown mineral drug for anesthesia in operations. They were also aware of the

*H. M. Bottcher, *Miracle Drugs* (Hienemann, London, 1963).

relationship between the nervous system and the movements of our limbs, and therefore understood the causes of paralysis. The Smith Papyrus contains forty-eight clinical cases. The ancient peoples of the Nile practiced hygiene, and, generally speaking, their medicine was far superior to that practiced so much later in Europe during the Middle Ages—yet another example of the withering of knowledge.

The physicians of the land of the pyramids were aware of the functions of the heart and arteries and how to count the pulse. Imhotep (4500 B.C.), the architect of Zoser pyramid, is considered to be the first recorded physician in history.

Ancient India possessed advanced medical knowledge. Her doctors knew about metabolism, the circulatory system, genetics, and the nervous system as well as the transmission of specific characteristics by heredity. Vedic physicians understood medical ways to counteract the effects of poison gas, performed Caesarean sections and brain operations, and used anesthetics.

Sushruta (fifth century B.C.) listed the diagnosis of 1,120 diseases. He described 121 surgical instruments and was the first to experiment in plastic surgery.

The *Sactya Grantham,* a Brahmin book compiled about 1500 B.C., contains the following passage giving instructions on smallpox vaccination: "Take on the tip of a knife the contents of the inflammation, inject it into the arm of a man, mixing it with his blood. A fever will follow but the malady will pass very easily and will create no complications." Edward Jenner (1749-1823) is credited with the discovery of vaccination but it appears that ancient India has prior claim!

The United Kingdom and other countries have medical-aid programs supported by the state. But the physicians of the Inca Empire and the Land of the Pharaohs also received their remuneration from the government and medical aid was free to all. Truly, there seems to be nothing new under the sun.

The Chinese Emperor Tsin-Shi (259-210 B.C.) possessed a

"magic mirror" which could "illuminate the bones of the body." X ray in ancient China? It was located in the palace of Hien-Yang in Shensi in 206 B.C. When a patient stood before this rectangular mirror, which was 1.76 by 1.22 meters in size, the image seemed to be reversed but all the organs and bones were visible exactly as on our fluoroscopes. That mirror was used for the very same purpose—to diagnose disease.

It is little known that a Chinese surgeon by the name of Hua T'o carried out operations under anesthetics over eighteen centuries ago. The chronicle *Hou Han Shu* of the later Han Dynasty (A.D. 25-220) reminds one of a report from a modern medical journal:

> He first made the patient swallow hemp-bubble-powder mixed with wine, and as soon as intoxication and unconsciousness supervened, he made an incision in the belly or the back and cut out any morbid growth. If the stomach or intestine was the part affected, he thoroughly cleansed these organs after the use of the knife, and removed the contaminating matter which had caused the infection. He would then stitch up the wound, and apply a marvellous ointment which caused it to heal in four or five days, and within a month the patient was completely restored to health.

The Lester Institute of Shanghai, founded by a British magnate in the thirties, has established scientific bases for old Chinese remedies. Every medicine, even as odd as donkey's skin, dog's brains, sheep's eyeballs, pig's liver, or seaweeds, has been found by Dr. Bernard Reed to possess a chemical reason for its effectiveness.

While blood transfusion was introduced into Western medicine in the seventeenth century, it has been practiced by the Australian aborigine for thousands of years. Our method is similar to one that he uses in that he takes blood either from a vein in the middle of the arm or from one in the inner arm by means of a hollow reed. Blood transference is also done by mouth but the technique of this method, though shown to various investigators, remains unfathomable.

Seemingly, the Australian medicine man is still heir to ancient knowledge. He is perfectly aware of the proper vein from which the blood should be taken. Uncannily, he also chooses the fitting donor. Blood transfusion is practiced not only in critical cases of injury and illness but also to give vitality to the aged.

When threatened by an impending drought or other calamity with the menace of food shortage, the aborigines have used oral contraceptives for centuries. Resin from a particular plant is rolled into pills to be taken by women.

Not only are these facts astonishing but it is also a pity that because of detribalization and lack of interest on the part of the medical profession, the herbal medicines of the Australian aborigines are almost forgotten. The natives do not cultivate the medicinal plants any longer and are in the process of losing their valuable heritage.*

*The Sydney Sun, August 8, 1969.

7

From Temples and Forums to Atomic Reactors

The so-called *Emerald Tables of Hermes* are of great interest to the student of the history of science. Although often considered as a document from the Middle Ages, its style and a total absence of medieval alchemical terms raises the possibility of its more ancient origin. Actually, on the basis of his research Dr. Sigismund Bacstrom, an eighteenth-century scholar, traced the *Emerald Tables* to about 2500 B.C.

"What is above is like what is below, and what is below is like what is above to effect the wonders of one and the same work," reads the opening sentence of the *Tables*. These words can be interpreted as the mirrorlike similarity between the world of the atom, with electrons whirling around protons as planets around the suns, and the macrocosm of stars and galaxies.

This idea of the oneness of the universe and the unity of matter is stressed again in another passage: "All things owe their existence to the Only One, so all things owe their origin to the One Only Thing.

"Separate the earth from the fire, the subtle from the gross, carefully and skillfully. This substance ascends from the earth to the sky, and descends again on the earth—and thus the superior and the inferior are increased in power." This paragraph might well be interpreted as the process of splitting the atom and the dangers connected with it.

"This is the potent power of all forces for it will overcome all that is fine and penetrate all that is coarse because in this manner was the world created," says another paragraph in the *Emerald Tables*. It indicates the belief of the ancients in the vibratory character of matter and the waves and rays which penetrate all substances.

Democritus was the first to formulate the atomic theory. Anticipating the views of modern physicists, he said almost 2,500 years ago: "In reality there is nothing but atoms and space." Moschus the Phoenician communicated to the Greek philosopher this primordial knowledge, and in fact, Moschus' concept of the structure of the atom was nearer to the truth because he emphasized its divisibility. His version of the atomic theory is being corroborated as new atomic particles are discovered all the time.

Greek philosophers claimed that there was no distinction in kind between the stellar bodies and the earth. The teaching of Hermes must have been accepted as an axiom by the Hellenic thinkers.

Leucippus (fifth century B.C.) as well as Epicurus (341-270 B.C.) also favored the atomic theory. Lucretius (first century B.C.), a Roman scholar, wrote about atoms "rushing everlastingly throughout all space." They undergo "myriad changes under the disturbing impact of collisions." It is impossible to see the atoms because they are too small, he asserted. These classic writers and philosophers command respect and admiration for their advanced thinking as they had anticipated modern science and have contributed to its development. But we still do not know *what* led them to believe in invisible atoms.

73

In his *On the Nature of the Universe* Lucretius expresses an opinion that "there can be no center in infinity." This thesis is the cornerstone of the Theory of Relativity of Einstein. Heraclitus (fifth century B.C.) must have likewise had relativist ideas because once he said: "The way up and the way down are one and the same."

Zeno of Elea (fifth century B.C.) demonstrated the relativity of motion and time by his paradoxical problems. "If the flying arrow is at every instant of its flight at rest in a space equal to its length, when does it move?" he asked. In his famous problem of the chariots Zeno even attempted to prove the time shrinkage of bodies in motion which Einstein dealt with more fully in his formulas.

Nicolaus, Cardinal of Cusa, a fifteenth-century scholar, wrote of a "universe without a center," thus giving another preview of the Theory of Relativity.

Lao Tse (sixth to fifth century B.C.), the founder of Taoism, taught that everything in the universe is made according to a natural law, or Tao, which controls the world. All creation is the result of the interplay of two cosmic principles—the male Yan and the female Yin—promulgated Lao Tse. Scientifically this is true because positive and negative charges in the nuclear world determine all manifestations in nature.

Ancient sages realized the dangers of revealing knowledge to those who could use it for destructive aims. "It would be the greatest of sins to disclose the mysteries of your art to soldiers," wrote a Chinese alchemist a thousand years ago. Are modern nuclear alchemists guilty of this sin?

The atomic structure of matter is mentioned in the Brahmin treatises *Vaisesika* and *Nyaya*. The *Yoga Vasishta* says: "There are vast worlds within the hollows of each atom, multifarious as the specks in a sunbeam."

The Indian sage Uluka proposed a hypothesis over 2,500 years ago that all material objects were made of *paramanu,* or seeds of matter. He was then nicknamed *Kanada,* or the swallower of grains.

The sacred writings of ancient India contain descriptions of weapons which resembled atomic bombs. The *Mausola Parva* speaks of a thunderbolt—"a gigantic messenger of death" —which reduced to ashes whole armies and caused the hair and nails of the survivors to fall out. Pottery broke without any cause and the birds turned white. After a few hours all foodstuffs were poisoned. The ghastly picture of Hiroshima comes to mind when one reads this ancient text from India.

"A blazing missile possessed of the radiance of smokeless fire was discharged. A thick gloom suddenly encompassed the heavens. Clouds roared into the higher air, showering blood. The world, scorched by the heat of that weapon, seemed to be in fever," thus describes the *Drona Parva* a page of the unknown past of mankind. One can almost visualize the mushroom cloud of an atomic bomb explosion and atomic radiation. Another passage compares the detonation with a flare-up of *ten thousand suns.*

The physicist Frederick Soddy evidently did not take these ancient records as fable. In the *Interpretation of Radium* (1909) he wrote these lines: "Can we not read into them some justification for the belief that some former forgotten race of men attained not only to the knowledge we have so recently won, but also to the power that is not yet ours?" When Dr. Soddy wrote the book, the atom-box of Pandora had not yet been opened.

A radioactive skeleton has been found in India. Its radioactivity was fifty times above the normal level.* Perhaps the Sanskrit texts about atomic warfare in protohistory are true.

The surface of the Gobi Desert near Lob Nor Lake is covered with vitreous sand which is the result of Red China's atomic tests. But the desert has certain areas of similar glassy

*A. Gorbovsky, *Riddles of Ancient History* (Moscow, 1966).

sand which had been present for thousands of years before Chairman Mao! What was the source of heat which melted that sand in prehistory?

The Brahmin books contain a curious division of time. For instance, the *Siddhanta-Ciromani* subdivides the hour until it arrives at the final unit—*truti,* equivalent to 0.33750 of a second. Sanskrit scholars have no idea why such a small fraction of a second was necessary at all in antiquity. And no one knows how it could have been measured without precision instruments.

According to Pundit Kaniah Yogi of Ambattur, Madras, whom I met in India in 1966, the original time measurement of the Brahmins was sexagesimal, and he quoted the *Brihath Sathaka* and other Sanskrit sources. In ancient times the day was divided into 60 *kala,* each equal to 24 minutes, subdivided into 60 *vikala,* each equivalent to 24 seconds. Then followed a further sixty-fold subdivision of time into *para, tatpara, vitatpara, ima,* and finally, *kashta*—or 1/300,000,000 of a second. The Hindus have never been in a hurry and one wonders what use the Brahmins made of these fractions of the microsecond. While in India the author was told that the learned Brahmins were obliged to preserve this tradition from hoary antiquity but they themselves did not understand it.

Is this reckoning of time a folk memory from a highly technological civilization? Without sensitive instruments the *kashta* would be absolutely meaningless. It is significant that the *kashta,* or 3×10^{-8} second, is very close to the life-spans of certain mesons and hyperons. This fact supports the bold hypothesis that the science of nuclear physics is not new.

The *Varahamira Table,* dated B.C. 550, indicates even the size of the atom. The mathematical figure is fairly comparable with the actual size of the hydrogen atom. It appears fantastic that this ancient science recognized the atomic structure of matter and realized how small is its ultimate particle. Nothing of this kind has ever been attempted in the West until the twentieth century.

Philolaus (fifth century B.C.) had a strange notion about an "antichthon," or "antiearth"—an invisible body in our solar system. It is only recently that the concept of antimatter, antiworld, and antiplanets has been introduced into science. In nuclear physics the positron is now known as an electron traveling from the future into the past. This time-direction reversal in the atomic world is a new discovery. But Plato wrote in the *Statesman* about an oscillating universe periodically reversing its time-arrow and sometimes moving from the future into the past. We know now that atomic particles can travel backward in time, but it seems that the idea was not unfamiliar to the great Plato.

While atomic knowledge in ancient times was fragmentary in character, we cannot say the same thing about astronomy. With its deep roots and constant practice over a period of millennia, the science of the stars reached a high level in antique times.

8

Sages Under the Heavenly Vault

The realization that the ancients somehow knew about the infinitely small appears to be impossible to us. There is also evidence which points to their knowledge of the infinitely great. No one can tell how the people of antiquity were able to obtain this information without advanced precision instruments which they apparently did not have.

This contradiction between poor instruments and rich knowledge has puzzled many a scientist and thinker. There are references to the solar parallax in classical sources. In modern times the first observation of the parallax of the sun was made by William Gascoigne in 1670 by means of a wire netting placed in front of the telescopic lens. But the sages of antiquity were not known to have had telescopes. How did they discover the solar parallax? To observe the apparent displacement of the sun amidst the stars because of the movement of the earth on its orbit requires advanced instrumentation.

How did the ancients know that the orbit of the earth was not round but elliptical? How did they arrive at the conclusion that the plane of the terrestrial orbit did not coincide with the plane of the earth's equator? Plutarch cites Aristarchus (third century B.C.) in introducing this subject: "The earth revolves in *an oblique circle* while it rotates at the same time about its own axis." This mystery of the history of astronomy was noted by the celebrated astronomers J. S. Bailly in 1781 and K. Gauss in 1819 and was mentioned in their monumental works.

In the *Timaeus* of Plato written about 2,400 years ago, the philosopher gives a dialogue between a high priest of Egypt and Solon, the lawgiver of Greece. A curious fact emerges from it—the sages of the land of the pyramids were aware of asteroids in space and their occasional collisions with the earth. Astronomy tells us today that a huge meteorite hit Arizona about 50,000 years ago producing a devastating explosion. The Barringer Crater, 1.60 kilometers wide, is a mark on the face of our planet caused by that meteorite. The 3.22-kilometer Chubb Crater in Canada is the site of another astronomical accident—the fall of a bolide some 4,000 years ago. The force of impact was equivalent to 200,000,000-megaton atom bombs. The earth has scars of this type in Saudi Arabia, Australia, and Africa, while the moon seems to be thoroughly pockmarked with them.

To collate data of unusual happenings in nature in the course of thousands of years and then evaluate them correctly can be done only by men of science. The sages of Egypt earned themselves such titles.

The old Egyptian priest called Solon's attention to the Greek legend of the fall of Phaethon and explained to him what it really meant: "Now this has the form of a myth but really signifies a deviation from their courses of the bodies moving around the earth and in the heavens, and a great conflagration of things upon the earth recurring at long intervals of time." Can anything be clearer? The sage alludes

79

to asteroids in space and their accidental crashes on our planet, causing explosions.

The Academy of Sciences of France made a statement in writing 170 years ago which showed its disagreement with the views held by the wise men of the land of the Nile: "In our enlightened age there can still be people so superstitious as to believe stones fall from the sky." This is but another example of the periodic triumphs of ignorance even in an "enlightened age."

It might be inferred that a scientific legacy has existed for thousands of years, and in spite of wars, famine, plagues, and other calamities which often destroyed whole civilizations, this age-old science was passed on from one generation to the next.

"Mosques fall, palaces crumble into dust but knowledge remains," said Ulug Beg, the great Uzbek astronomer of the fifteenth century. For these defiant words the scholar was ordered to go on a pilgrimage to Mecca. He never reached Arabia because the ruler's agents murdered Ulug Beg on the road. The names of his assassins have been forgotten but after five centuries Ulug Beg's astronomical tables are still used because of their accuracy.

Long before mosques and palaces there lived a strong astronomical tradition—even in caves. Stone cavings at Pierres Folles, La Filouzière, Vendie, and Brittany have been identified as prehistoric astronomical charts. The constellations of Ursa Major, Ursa Minor, and the Pleiades are represented by clusters of small hollows in the rock. Since astronomy would seem to have no practical use for hunters dwelling in caves, what was it that stirred their interest in stargazing?

Thousands of Ice Age notational sequences—that is, vertical markings, lines, and dots, painted and engraved on stone or bone—are scattered from Spain (Canchal de Mahoma and Abri de las Vinas) to the Ukraine (Gontzi).

In the Upper Paleolithic period, from about 35,000 B.C. to

8,000 B.C., a great number of these markings in the so-called Azilian, Magdalenian, and Aurignacian cultures are to be found. The fact that one type of linear art had existed in prehistory in an unbroken line for some 30,000 years is in itself very significant.

The American scientist Alexander Marshack made the discovery that these notations on rock or bone are records of observations of the moon, made for calendric purposes.* To find complex memoranda of lunar studies in prehistory was astounding to science. Marshack believes that his discovery "entails a revaluation of the origins of the human culture." He also states that because of this evidence "an earlier basic astronomical skill and tradition existed." This discovery is revolutionary in nature, calling for a review of the intellectual powers of man in the last Glacial Age.

Although this prehistoric calendrical system may have no connection with the Scandinavian runes, yet both are arranged as notches on a foot rule. The runic calendar appeared in the north of Europe about 2,000 years ago, and its use was abandoned only in the early part of the nineteenth century. In fact, runic rulers are permanent calendars and can be used today! To develop a chronology requires a knowledge of astronomy and mathematics, accumulated during the course of long ages. The runic calendar of the Baltic basin may be a grandchild of the prehistoric notational system. Dr. L. E. Maistrov of the USSR has the opinion that the runic calendar is based on the solar cycle of 28 years. The beginning of this calendrical system goes back to the year 4713 B.C. Although the runic calendar may not be as old as that, its starting point was placed far back in time.

The first calendar in Egypt began with the earliest recorded date, 4241 B.C. Egyptian star charts occur as early as 3500 B.C., indicating a systematic study of astronomy. The Egyptians were aware that Mercury and Venus were closer to

*Science (U.S.A.), November 6, 1964.

the sun than were the earth, Mars, Jupiter, and Saturn.

Observations of the motions of Venus, Mars, and Jupiter were recorded in cuneiform writing by the priests of Babylon almost 4,000 years ago. The astronomy of Mesopotamia was more advanced and more accurate than that of Egypt, inasmuch as the Babylonian priests were able to forecast eclipses.

The ancient inhabitants of England were even more proficient in astronomy than the priests of Egypt or Sumer. The computations of the Stonehenge alignments made by Professor Gerald S. Hawkins have disclosed a precise knowledge of the solstices and the equinoxes, and the ability to predict eclipses by the builders of those megaliths around 2000 B.C.* The complexity of the Stonehenge astronomical tradition indicates a development of some thousands of years. Did this science develop locally or was it imported from another center of civilization?

The first encyclopedists lived in ancient Greece. They had not only collected, classified, and assimilated the science of the older civilizations of Egypt and Sumer but had also drawn their own brilliant conclusions.

"The earth is round and it revolves around the sun," said Anaximander (c. 610-547 B.C.). "The earth is a globe," taught Pythagoras to his disciples in Crotona in the sixth century B.C.

Aristarchus of Samos (310-230 B.C.) affirmed that the earth traveled in an orbit around the sun, rotating on its axis at the same time. He even added that all the planets moved around the sun.

"The earth spins on its axis once in twenty-four hours," said Heraclides of Pontus in the fourth century B.C. Seleucus of Erythrea (second century B.C.) also spoke about the rotation of the earth and its orbit around the sun.

"I want to find out the size of this earth," said Eratosthenes (c. 276-194 B.C.), the custodian of the Alexandrian Library.

*G. S. Hawkins, *Stonehenge Decoded* (Souvenir Press, London, 1965).

He noticed that due south at Syene the sun was directly overhead on midsummer's day and seven degrees from vertical at Alexandria on the same day. With the aid of geometry he obtained a figure for the circumference of the earth, and then one for its diameter. Surprisingly, there was a discrepancy of merely 80 kilometers between his estimated figure for the polar diameter and that accepted by our modern astronomy.

When Megasthenes, the Greek ambassador to India, introduced the subject of astronomy during his audience with King Chandragupta Maurya in 302 B.C., the latter declared: "Our Brahmins believe the earth to be a sphere."

The ancient book *Surya Siddhanta* contains reasonably accurate calculations of the diameter of the earth and its distance from the moon. The *Rig Veda,* the sacred book of India, contains a curious passage concerning the *"three earths"*—one within the other. The earth does have *three* thick zones—the inner core, outer core and mantle, plus a very thin crust. It is only with the advancement of our science and the perfection of instruments that we have discovered the veracity of the *Rig Veda.*

Knowledge is power and the priests of India, Babylon, Egypt, and Mexico wanted to keep it. No wonder the sixth chapter of the *Surya Siddhanta* insists: "This mystery of the gods is not to be imparted indiscriminately." This old law had been so rigid in India that if a man of a lower caste tried to listen to the *Vedas* being read, molten lead was poured into his ears as punishment. The British finally put an end to this cruel custom early in the nineteenth century.

That the astronomers of old were required to be competent and industrious can be seen from an episode related in the *Book of Shi Ching* (*Book of Odes*). During the reign of Emperor Yao of China (*c.* 2500 B.C.) the two official astronomers Hi and Ho got into the bad habit of drinking too much hot rice wine and one day failed to make an announcement of a coming eclipse. But the law was strict where their duties were concerned: "If an eclipse took place before the estimated time, the astronomers were to be killed without respite." Now if the phenomenon occurred after the predicted date they "were to be slain without reprieve." The end of the story is sad. Hi Ho! the stargazers were dispatched to the stars. But later chronicles of China, such as that of Chou in the twelfth century B.C., contain precise astronomical forecasts of the eclipses of the moon.

Nan-chi Hsien-weng, a hero of the Chinese pantheon, had the enigmatic title of "Ancient Immortal of the South Pole." According to tradition, he helped the general Chiang Tzu-Ya in the year 1122 B.C. It appears that over 3,000 years ago the scholars of China had a correct concept of the spherical shape of the earth when they spoke of the South Pole.

"The earth is an egg," said Chang Heng (A.D. 78-139) and explained that its axis pointed to the Polar Star.

We have evidence stretching back for thousands of years which is unanimous in assuring us that some ancient thinkers had a perfectly scientific concept of the earth in space.

Before embarking on his historic voyage Columbus made a study of all classical sources regarding the shape of the earth and the possibility of reaching the East by taking a westward route. In a letter preserved in Madrid the discoverer of America made the curious statement that the earth was *slightly pear shaped*. Satellites have recently disclosed that our planet is slightly pear shaped. How was Christopher Columbus aware of this fact unless he had found it in an ancient text?

Let us speak now about the moon, which has received so much publicity since the beginning of Apollo missions. The *Surya Siddhanta* contains a passage relating to "the radiant sun which supplies the moon with light rays," evidently alluding to the reflected light of the moon.

Parmenides made a definite statement about the moon in

84

the sixth century B.C.: "It illuminates the nights with borrowed light." This is an obvious reference to the reflection of the sun's rays from the lunar surface. Empedocles (494-434 B.C.) held the same opinion: "The moon circles round the earth—a borrowed light."

Twenty-five centuries before our lunar exploration Democritus exclaimed: "Those markings on the moon? They are shadows from high mountains and deep valleys."

"It is the moon that darkens the sun during an eclipse," said Anaxagoras 2,500 years ago. And he was also the first to explain that during a lunar eclipse it is the earth's shadow that falls on the moon.

The words of Plutarch regarding the moon were indeed prophetic: "If you regard her as a star or a certain divine and heavenly body, I am afraid she will prove deformed and foul," he declared. Lunar photographs and television pictures do show dreary wastelands.

An ancient Brahmin tradition teaches that "Lunar Pitris," or the patriarchs, created all life on this planet after their descent from the moon. The Sanskrit texts always connect the Pitris with the moon and the kingdom of the dead, which would seem to imply that the moon is older than the earth. The seven ages of the ancients were in some way connected with the planets. The moon stood for the cradle of life. This belief that the moon had its day before the earth has no logical explanation. In Mayan art the moon god is depicted as an old man with a conch shell. The moon goddess of ancient Mexico, Ixchel, was addressed as the Grandmother. "In the religion of many primitive peoples, the moon is considered to be the first man who died," states *Encyclopaedia Britannica.*

These ancient beliefs concerning the great age of the moon are now corroborated by the mineral samples brought from the moon by Apollo 11. The rocks from the moon's surface were dated 4.6 milliard years whereas the oldest minerals on this planet are 3.3 milliard years old.

The ancients saw the connection that existed between the tides and the moon. Seleucus, an astronomer of Babylon, correctly explained the tides of the seas by lunar attraction. The sages of China also had no doubts that it was the pull of the moon that was responsible for the rise of the sea level.

Julius Caesar was a better general than scholar but even he wrote that when the moon is full the tides are high, and he waited for the high spring tides to land in England. That was 2,000 years ago. When in the sixteenth century the great German astronomer Johann Kepler announced his theory that the tides were caused by the moon, he was severely censured. Kepler could not afford to argue because a relative of his was burned as a witch before his eyes, while his mother died in prison in chains. This historical episode demonstrates once more the obscuration of science and the persecution of men who were trying to resurrect the knowledge of antiquity.

In the tenth century the Arab astronomer Abul Wafa wrote about the "variation of the moon." As the path of the moon is an ellipse, our satellite is 3,219 kilometers closer to the earth at new-moon point, and 2,575 kilometers farther at last-quarter point. This discovery is generally attributed to Tycho de Brahe (1546-1601). However, the treatise of his Arab colleague written six centuries earlier mentions this irregular wobbling of the moon.

Since a chronometer is required for such measurements, strictly speaking, without a good clock—which Abul Wafa did not have—it would have been impossible to observe the lunar variation. The controversy is still raging—who discovered the variation of the moon?

From Luna our road leads to the sun. "The sun is a vast mass of incandescent metal," boldly declared Anaxagoras 2,500 years ago. But the pious citizens of Athens believed otherwise—the sun was the throne of Apollo. Anaxagoras said the right thing at the wrong time and was exiled. And about the same time Democritus, of atomic-theory fame, postulated that the sun is of immense size.

Before Galileo no one knew anything about sunspots; nor was the sun, as a perfect, divine, stellar body, supposed to have any spots. But 2,000 years ago the Chinese made astronomical records of sunspots.

The *Vishnu Purana* reads: "The sun is always in one and the same place." This sentence alludes to the apparent motion of the sun from east to west and suggests that it is the earth which rotates.

Ancient Mexico had an incredibly high degree of astronomical knowledge. The actual figure for the duration of the year is 365.2422 days, according to present-day astronomy. Our Gregorian calendar takes it as 365.2425. But the Mayas computed the length of the year as 365.2420, the closest to the sidereal figure. In other words, the ancient Central American Indians had a more precise calendar than we in this Age of Science!

The Copán Mayas estimated the duration of the lunar month as 29.53020 days, and the Palenque Mayas as 29.53086. According to astronomy, the figure is 29.53059 days. How did the Mayas get their results without chronometers and all the precision instruments we possess today? Actually, the correct figure is just halfway between the estimations of Copán and Palenque.

The Stele I of El Castillo in Santa Lucia Cotzumahualpa in Guatemala represents the transit of Venus on the solar disk on November 25, 416. This discovery was made by C. A. Burland, who reported it to the International Congress of Americanists in Paris in 1956. While at the Congress he stated that "The Cotzumahualpan astronomers were serious and accurate scientists." Now to reach an advanced knowledge of astronomy of this kind, science requires many centuries of continuous and uninterrupted development. Quite possibly we may be dating incorrectly the beginning of civilization in Central America.

In the British Museum there are Babylonian inscriptions which speak of the "horns of Ishtar" (Venus), or the crescent of the planet. However, this crescent is visible only through a telescope. Although the German astronomer K. Gauss wrote early in the nineteenth century about the ability of his mother to see the phases of Venus with the naked eye, no other such historical cases had been known.

The first astronomical observation of Venusian phases was made by Galileo in 1610. He left the following anagram to claim priority rights: *"Cynthiae figuras aemulator Mater Amorum,"* or "The Mother of Love (Venus) imitates the figures of Cynthia (the moon)."

Why was Ishtar, or Venus, called "Sister of the Moon" or the "Mighty Daughter of the Moon" by the Babylonians? Why not "Sister of Jupiter," which it resembles so much more in brilliance? Perhaps the explanation is that the scientific priesthood of Babylon had somehow known about the moonlike phases of Venus.

The Babylonian priests also recorded their observations of the four greater satellites of Jupiter—which, again, cannot be seen without a telescope. In alluding to this fact Professor George Rawlinson wrote: "There is said to be distinct evidence that they observed the four satellites of Jupiter and strong reason to believe that they were acquainted likewise with the seven satellites of Saturn."*

The discovery of the four moons of Jupiter was made by Galileo in 1610. The Saturnian satellites were first observed by Cassini, Huygens, Herschel, and Bond between 1655 and 1848. How could the Babylonians have known about them? Did the priest-astronomers of Babylon have superhuman

*G. Rawlinson, *The Five Great Monarchies of the Ancient Eastern World,* Vol. 3 (New York, 1880).

eyesight, telescopes—or a secret tradition from a lost civilization? Actually, there is a crystal disk in the British Museum found at Nineveh by Layard; at first it was considered to be a lens but it is not strong enough for astronomical work.

There is another unsolved case in the Scotland Yard of the history of science. The Dogons of Sudan have a strange tradition about the "dark companion of Sirius." This dim companion of the bright star Sirius is visible only in the most powerful telescopes, such as that of Mount Palomar.*

Only a few brief centuries ago the scholars and clerics of Europe believed in a stationary earth—the center of the universe—or even a flat earth with a firmament. The stars were holes in that firmament through which shone the light of paradise.

But in Greece in the fifth century B.C. Democritus said: "Space is filled with myriads of stars and the Milky Way is but a vast conglomeration of distant stars." It should be borne in mind that in Democritus' time no more than six thousand stars could be seen in the sky. By using logic and imagination he arrived at the correct picture of the universe which we have rediscovered only during the last 150 years.

Thales of Miletus (c. 640-546 B.C.) was another genius. He concluded that the stars were of the same substance as the earth. This idea of the universality of matter was buried in the Middle Ages and resurrected only yesterday.

"The distances which separate us from the stars are immeasurable," said Aristarchus of Samos twenty-three centuries ago.

"There are more planets than the ones we can see," taught Democritus. What gave him the idea that there were planets beyond Saturn? When Democritus was a young man, Anaximenes spoke of "nonluminous" companions of stars.

*M. Agrest, *Literatournaya Gazeta* (Moscow, 1963).

Surely he was not referring to planets in other solar systems. Or do we underestimate his intelligence and imagination?

Seneca (4 B.C.-A.D. 65) in *Natural Questions* shows deep insight in his speculation on astronomy and its future: "How many heavenly bodies revolve unseen by human eye! How many discoveries are reserved for the ages to come when our memory shall be no more." How right he was. Uranus, Neptune, and Pluto were discovered only in the last 200 years. And while only a few thousand stars were known in Seneca's time, millions are now listed in our star catalogues.

The Tanguts, a central Asian tribe whose city of Hara-Hoto was excavated in 1908, had a strange belief about the *eleven* luminaries—the sun, the moon, Mercury, Venus, Mars, Jupiter, Saturn, and the planets Tsi-Tsi, Ouebo, Rahu, and Ketu. While Rahu and Ketu are undoubtedly the ascending and descending nodes of the moon, borrowed from Hindu astronomy, the identity of Tsi-Tsi and Ouebo remains a mystery. Are they Uranus and Neptune?*

One of the most amazing things mentioned in ancient texts and legends is the notion of life in other worlds. To the legendary Orpheus, son of Apollo, is attributed this fragment: "Those innumerable souls, they fall from planet to planet and, in the abyss of space, lament the home they have forgotten." These words appear to speak of life on other planets.

Heraclitus (*c.* 540-475 B.C.) and all the disciples of Pythagoras (sixth century B.C.) considered each star as being the center of a planetary system. Democritus taught that worlds come into being and die. Only some of these worlds in the stars are suitable for life, he said.

Anaxagoras (500-428 B.C.), another Greek philosopher, also wrote about "other earths which produce the necessary sustenance for inhabitants."

Metrodotus of Lampsacus (third century B.C.) believed in

*E. I. Lubo-Lesnichenko and T. K. Shafranovskaya, *The Dead City of Hara-Hoto* (Moscow, 1968).

the plurality of populated worlds. He said that to call the earth the only inhabited world was as unwise as to assert that there was only one spike of grain growing in a vast field. Epicurus (341-270 B.C.) was likewise convinced that life was not confined to our planet alone. Lucretius (c. 98-55 B.C.), a Roman poet, wrote that "it is in the highest degree unlikely that this earth and sky is the only one to have been created."

According to Cicero (106-43 B.C.) the realm of heaven was peopled with a host of genii. This notion is remarkable as it approaches our present-day idea of inhabited worlds in space.

Was this a brilliant speculation or a heritage from some Golden Age of Science? If it were only a conjecture, why was it identical in countries so widely separated geographically as Mexico, China, Greece, India, Egypt, and Babylon?

Why did the Romans have the so-called Cumaean Prophecy? In his *Fourth Eclogue* Virgil recorded it: "Now a new race descends from the celestial realms." This line does not speak of ethereal creatures but of a new people coming from starry space. The origin of so astonishing a concept has yet to be explained, for we are referring back to the period of the reign of Emperor Augustus.

The *Vedas* of India are quite definite about "life on other celestial bodies far from the earth." Teng Mu, a scholar of the Sung Dynasty, summed up the views of ancient Chinese thinkers on the universality of life: "How unreasonable it would be to suppose that besides the earth and the sky which we can see there are no other skies and no other earths."

Now let us turn our attention to comets. From 204 B.C. Chinese astronomers kept records of every appearance of Halley's Comet. In 11 B.C. they watched a comet for nine weeks, making descriptions of its changing shape exactly as do the astronomers of today.

"Comets move in orbits like the planets," wrote Seneca nineteen centuries ago.* Aristotle cites the Pythagoreans in

*Quaestiones Naturales, Book VII.

91

identifying comets as stellar bodies reappearing before long periods of time. This reasoning was magnificent because comets do not carry identification plates. On the authority of Apollonius Myndius it can be surmised that the doctrine came from Babylon, antedating Pythagoras by many a century.

In the second century A.D. the Roman historian Suetonius defined comets as "blazing stars which are thought by the ignorant to portend disaster to rulers."

But what happened fourteen centuries after Suetonius in that choicest portion of the earth—Europe—is incredible by its stupidity. The town council of Baden, Switzerland, issued a proclamation in January, 1681, when a "frightful comet" with a long tail appeared in the sky. "All are to attend Mass and Sermon every Sunday, abstain from playing and dancing, and evening drinking is to be on a modest scale," it said. Did almost all mankind bury its head in the sand after the noble Greeks, proud Romans, and intuitive Egyptians had made their exit?

Cosmology as a science began with Kant and Laplace only 200 years ago. However, the *Huai Nan Tzu* Book (*c*. 120 B.C.) as well as the *Lun Heng* written by Wang Chung (A.D. 82) outlined the crystallization of worlds by whirlpools of primary matter.

The ancient *Popul Vuh* of the Guatemalan Mayas thus describes the appearance of the world: "Like the mist, like a cloud, and like a cloud of dust was the creation." And here is a modern version of the same cosmogony: "The stage began with the precipitation of the dust specks of the central [equatorial] plane of the flattened cloud."* What was the source of Mayan cosmology? Was it the same that gave them the most precise calendar in the world?

---

*B. Levin, *The Origin of the Earth and Planets* (Moscow, 1958).

9

The Zodiac and the Music of the Spheres

How strange it is that the ancient Mayas call the constellation known to us as Scorpion by the same name? Orion, or the Hunter, of Babylon, Egypt, and Greece had a similar name in China—the Hunter of the Autumn Hunt. Our Aquarius is echoed in the Mexican god Tlaloc, the Ruler of the Rains.

But what is really puzzling is this—it is only by a stretch of imagination that one can find a connection between the figures the constellations are supposed to represent and the formations of stars. It looks as if early civilizations had access to older lists of constellation names which they adopted to identify the myriads of stars.

The Chinese zodiacal sign of the Sheep finds its replica in the Babylonian zodiac as Aries. The Ox sign of China finds its reflection in the West as Taurus. The Horse of Chinese astronomy is Sagittarius in Babylon and Egypt. Although the names are often identical, they sometimes do not refer to the same constellation.

The similarities in the constellation names in Central America and China are even more striking. The Aztec calendar has days of the Alligator, Snake, Rabbit, Dog, and Monkey. The Chinese-Tibetan calendar has the years of the Dragon, Snake, Rabbit, Dog, and Monkey as well!

These strange coincidences ought to be examined. We cannot but agree with the eminent scientist Giorgio de Santillana, who writes this about constellation names in *The Origins of Scientific Thought*: "They were repeated without question substantially the same from Mexico to Africa and Polynesia—and have remained with us to this day."

While the Pythagoreans were observing the constellations, a curious thing happened—some heard the Music of the Spheres. According to Pythagoras, all stars were alive and they harbored intelligences. He imagined the universe to be a cosmic lyre with numerous strings which gave out this Music of the Spheres. Around the same time the Taoists of China held the same concept: "Everything that is is in space, and everything that is in space has a sound."

Today we know that stars and planets are sources of radio emission. Is this what the Pythagoreans and the Taoists meant? Do our radio telescopes pick up this Music of the Spheres?

The musical scale was introduced by Pythagoras. By measuring the length of chords and listening to sounds from strings, he discovered a mathematical correlation.

His school postulated that the planets moved in an orderly manner and that their distances from the "central fire" were not unlike the intervals in the diatonic scale of music.

Pythagoras and his disciples were not far from the truth because planetary orbits are arranged in a certain mathematical order. According to Bode's Law, if the number 4 is added to 0, 3, 6, 12, 24, etc., and the result is divided by 10, we get the approximate distances of the planets from the sun, taking the distance from the earth to the sun as one unit.

Planet		Bode's Law	Actual Dist.
Mercury	$0 + 4 = 4 \div 10 =$	0.4	0.39
Venus	$3 + 4 = 7 \div 10 =$	0.7	0.72
Earth	$6 + 4 = 10 \div 10 =$	1.0	1.00
Mars	$12 + 4 = 16 \div 10 =$	1.6	1.52
etc.			

This is another example of how the ancients already knew what were later believed to be new discoveries in astronomy.

Down through the ages astronomy has been connected with astrology, or the art of evaluating the influence of heavenly bodies upon earth and man. The fossil sciences of alchemy and astrology have not been free from superstition and distortion. Astrology is defined by science as primitive astronomy. However, its attempt to foretell the future has been regarded by savants as a fallacy.

Only a few decades ago no scientist would have believed that sunspot activity could produce a devastating invasion of locusts in Southeast Asia. The idea of such a connection between sunspots and insects on earth would have seemed too ridiculous to be discussed on an academic level. However, observations made during the past decades have established a coincidence which does exist between the locust invasions and sunspot activity. Appropriate measures are taken now before the locusts arrive.*

Serious scientists are beginning to pay attention to this influence of forces in space upon phenomena on earth. In his *Mysteries of the Universe* W. R. Corliss writes: "Stranger still is the observation that sunspot maxima are roughly synchronized with the French and Russian revolutions, both world wars, and the Korean conflict. If there is some small truth in astrology, the thing to do is to explain this truth in scientific terms and strip away all the pretence."†

*T. Chestnov, *Are We Alone in the Universe?* (Moscow, 1968).
†W. R. Corliss, *Mysteries of the Universe* (Crowell, New York, 1967).

According to the Soviet astronomer R. P. Romanchuk, the so-called squares and conjunctions in astrology have a scientific basis. It is the positions of the sun, Jupiter, and Saturn that determine sunspot activity, he says, basing his conclusion on a chart which he drafted.

In the Russian science magazine *Znanie-Sila* (No. 12, 1967) A. Gangnus writes: "In ancient times astrologers attempted to predict the future by the respective positions of the planets. Who knows, this may not be so absurd. If the respective positions of the planets really influence the sun, then astronomical tables could become data for heliogeophysical and even for long-range climatic forecasts."

The Soviet astronaut A. A. Leonoff and Dr. V. I. Lebedeff write: "The number of car accidents increases four times on the second day after the solar flare-ups, as compared with the days when the sun is calm."* They also state that suicides increase four to five times above the normal rate during the periods of explosions on the sun.

These quotations from serious scientific sources establish coincidences between solar as well as planetary forces and events taking place on our earth. Astrology was built on this very assumption that the stars influence our lives. It appears therefore that some of the beliefs of astrologers may have had a scientific basis!

When Halley criticized Newton for considering astrology as a science, Sir Isaac Newton replied: "I have studied the subject, sir. You have not." If the divinatory character of astrology is ignored, one thing becomes strikingly clear—the sages of antiquity had a clear concept of stellar bodies emitting radiation. That idea alone is thoroughly scientific.

---

*A. A. Leonoff & V. I. Lebedeff, *Cognition of Distance and Time in Space* (Moscow, 1968).

10

Apes and Ages

At a time when even a schoolboy knows that the Sinanthropus, our apelike prehistoric patriarch, lived over 600,000 years ago, it is curious to recollect the words of an early-nineteenth-century luminary of science, Cuvier, who once declared that "Prehistoric men, physically different from the men of today, have never existed on earth." But they have, and their skeletons can now be seen in museums.

It is significant that while the Europeans of a century or two ago erroneously traced the origin of man and the universe to a date less than 6,000 years before their time, many thinkers of the ancient past had a truly scientific concept of the long evolution of man.

According to the Sanskrit *Book of Manu* (*c.* second century B.C.) the germ of life first appeared in water from the action of heat. Then it manifested itself as a mineral, a plant, an insect, a fish, a reptile, a mammal, and finally in the form of a man. Other Brahmin scriptures of great antiquity list the *Incarnations of Vishnu* in the following order: fish, tortoise, boar, lion-man, dwarf, man with an ax, Rama, and Krishna. Again we can recognize in this allegory a pre-Darwinian notion of evolution. The fish becomes a reptile, the mammal comes to take its place. Then giant and dwarf primates appear. The Gigantopithecus was five meters tall while the Pithecanthropus was short. The Cro-Magnon, the "man with an ax," was the true progenitor of modern man. Rama is a symbol of civilized man. Krishna stands for the future goal of mankind—cosmic man. These remarkable ideas of Indian sages antedate the Theory of Evolution by thousands of years.

"Man's procreator is a fish—living creatures came from water," said Anaximander in the sixth century B.C. Lucretius, a Roman poet of the first century B.C., drew a picture of the "survival of the fittest" in his poem *On Nature.*

It is quite evident that these clear-cut notions of evolution existed long before Lamarck and Darwin. Only 100 years ago Darwinists met a wall of opposition, ridicule, and fear. At a lecture on evolution given in those troublesome years the wife of Sir David Brewster, an eminent savant, fainted on hearing the facts which her tender ears could not take. Even until recent years certain states in America have had a law prohibiting the teaching of evolution. Actually, there is nothing offensive in the Theory of Evolution—a cosmic process of growth from the low forms of life toward the superior holds a promise of a greater future for man. And who knows, perhaps the ancient Mayas were right. According to their sacred book the *Popul Vuh,* the monkey is a descendant of early man.

A comparison of scientific knowledge prevalent over 2,000 years ago and the beliefs current in the past 300 years forces upon us the conclusion that the ancients surpassed our ancestors in the interpretation of the phenomena they had observed.

The people of antiquity believed in the tremendous age of the world and mankind, which they estimated in tens of thousands and even millions of years. To the European of Napoleonic times, the earth and man were created by God only several thousand years ago. However, the Asiatics had different views.

The Brahmins of India calculated the duration of the universe, or the Day of Brahma, to be 4.32 billion years. The Druses of Lebanon set the beginning of creation at 3.43 billion years. The present age of the earth is considered to be about 4.6 billion years, whereas that of the crust is 3.3 billion years. There are strange parallels between these figures. What is really extraordinary is the pundits' time reckoning in milliards of years—cosmic chronology of this type was unknown until this century.

According to Simplicius (sixth century A.D.) ancient Egyptians kept records of astronomical observations for 630,000 years. The archives of Babylon were 470,000 years old, wrote Cicero with a remark that he did not believe this claim. Hipparchus (c. 190-125 B.C.) mentioned Assyrian chronicles stretching back for 270,000 years.

The Egyptian priests told Herodotus in the fifth century B.C. that the sun had not always risen where it rose then. This implied that they had kept records of the precession of equinoxes, covering at least 26,000 years.

The Greek historian Diogenes Laertius (third century A.D.) claimed that the astronomical records of Egyptian priests began in 49,219 B.C. He also referred to their registers of 373 solar and 832 lunar eclipses, which would involve a period of approximately 10,000 years.

The Byzantine historian George Syncellus said that the chroniclers of the pharaohs had recorded all events for 36,525 years. Martianus Capella (fifth century A.D.) wrote that the Egyptian sages had secretly studied astronomy for over 40,000 years before they imparted their knowledge to the world.

The first dynasty after the Deluge was traced by Babylonian priests to a date 24,150 years before their time.

According to Codex Vaticanus A-3738, the Mayas had kept their calendrical system since 18,612 B.C.

Herodotus places the reign of Osiris at about 15,500 B.C. from the information given to him by the priests of the land of the Nile. He made the remark that they were quite certain about the exactitude of the date.

The lunar calendar of Babylon and the solar calendar of Egypt coincided in the year 11,542 B.C. The calendrical computations of India began with the year 11,652 B.C.

According to Plato, the Egyptian priests fixed the date of the sinking of Atlantis at 9850 B.C. while the Zoroastrian books set the "beginning of time" at 9600 B.C.

That these dates are correct can be questioned. But we cannot escape the conclusion that the ancients were much closer to truth than the scholars and clerics of one and a half centuries ago who thought that the world had been created in 4004 B.C. according to the Biblical chronological study of Bishop Ussher.

The universe of the Brahmins was almost as old as that of modern science. The chronicles of the Mayas, the Egyptians, and the Babylonians went farther back in time than our history. In view of what our science has yet to learn, it would be presumptuous to accuse them of exaggeration.

The mental horizons of the peoples of antiquity were vast and we are only beginning to see today what they perceived yesterday.

The priests of Babylon and Egypt believed that man was civilized 500,000 years ago. They kept historical and astronomical records in their archives, as Simplicius and Cicero tell us. We can smile at these claims and give civilization 5,000 years to progress from the chariot to the automobile, from bows and arrows to the atomic bomb, from the boat to the spaceship.

Although legitimately drawn from the palaeontological evidence available today, certain inferences of anthropology are questionable, as we shall later see. According to anthropology, anthropoid apes appeared about 2,000,000 years ago. They were neither men nor apes. There is a possibility that both modern man and the present-day ape had a common ancestor.

Accordingly, if this period of 2,000,000 years, representing the life-span of man, is likened to a year, then the Australopithecus appeared on July 1, the Pithecanthropus came on October 14, the Neanderthal on Christmas Day, and the Cro-Magnon on December 27, 28, 29, and 30. Today is December 31 on this scale, and it has been about 5,500 years long.

Primitive man began to make tools between July and September of this Great Year, and early in the second week of December he discovered fire. For eleven months of this Great Year of Evolution the ancestor of man was slowly detaching himself from the animal kingdom by adopting an erect posture and developing a larger brain.

Our immediate progenitor is the Cro-Magnon. He was a 6-footer (1.8 meters), intelligent and good-looking. He originated in the last Ice Age about 35,000 years ago and had a continuous existence until the dawn of history when he became the forefather of modern man.

The Neanderthal of Europe was nothing like him. He was only 1.65 meters tall, with short muscular limbs and a broad chest, weighing about 82 kilograms. This prehistoric man had a very small forehead, almost no chin, and, in comparison with the Cro-Magnon, was ugly. During the earliest period of his existence the Cro-Magnon was contemporaneous with the Neanderthal, whom he removed from the European scene by his strength and intelligence. The Neanderthal was not the grandfather of the Cro-Magnon, although occasional crossings of races did take place to the benefit of the more primitive, creating a mixed Neanderthal type.

Evolution labored for hundreds of thousands of years to

produce the Neanderthal out of the primates. If so long a period of time was necessary to evolve this foreheadless, chinless, thick-necked, stocky creature, how could the more evolved Cro-Magnon have developed in the space of a few thousand years? Here is his portrait—he dressed himself in skin clothes which were sewn and embroidered. He carved mammoth bone, painted beautiful pictures on rock, and kept calendars by watching the moon. He even had art schools.

Our planet has had unexpected glacial epochs of different durations. The last one ended about 12,000 years ago. But there were interglacial periods before the one we live in today. A warm climate prevailed about 150,000 years ago during which civilization might have been born, flourished, and then died in an avalanche of ice and ocean waves. The Cro-Magnon could have been a survivor from this Garden of Eden. This hypothesis would account for the large brain and high forehead of the Cro-Magnon. He may have brought with him hereditary traits from a former race, just as we ourselves carry his genes.

Now, using the same comparative scale, let us divide the day we are living—December 3—into twelve hours, from 6 a.m. to 6 p.m., from sunrise to sunset. Today at 7 o'clock in the morning we discover bronze, writing, and the wheel. At 8 we begin to build cities. Shortly before 11 we learn how to melt and forge iron. Between 1 and 2 in the afternoon the Greek forefathers meditate on the nature of the universe—from the atom to space travel. About 4:30 p.m. we emerge from an historical siesta—the Dark Ages—and begin to develop the scientific legacy of Greece. At 5 in the afternoon, explorers sail the oceans, opening new continents. At sunset we steal the Promethean fire of the atom and then soar to the moon. Everything happens in the last hour of the last day of the Great Year.

If all the above is correct, then the story of man is un- paralleled in evolution. It took the horse 60,000,000 years to become what it is now. The ancestor of the ant lived 150,000,000 years ago, and his descendants have changed but little.

There is something strangely unrealistic in the picture of a tree-climbing animal's becoming in 2,000,000 years a biped who can make machines to sail on water, roll on land, or fly in the air or interplanetary space, while his retarded cousins still jump from tree to tree. It is difficult to believe that man's history is so short while that of a horse is thirty times longer. could not have displayed his artistic talents without heredity from another cycle of civilization of which we know nothing. Neither could we have reached the moon without the biological legacy from the Cro-Magnon.

Only material evidence from protohistory can make this speculation truly scientific. But there is a great deal to suggest that the evolutionary road of mankind is much longer than it is considered at present. The discovery of a man-type skeleton in Tuscany in 1958 by Dr. J. Hurzeler and Dr. H. de Terra in a 10,000,000-year Miocene stratum lends strength to the author's theory of the ancientness of man.

Or has our growth been accelerated by another galactic civilization, millions of years older than our own? "Maybe we are property," mused Charles Fort—property of a cosmic supercivilization breeding gods out of monkeys?

11

The Celestial Comedy

In 1877 Asaph Hall, the director of the Naval Observatory in Washington, discovered the two small moons of Mars—Phobos and Deimos.

Curiously enough, the fifteenth book of the *Iliad* alludes to the fact that the god Mars had two companions—Phobos and Deimos. Was an ancient tradition concerning the Martian satellites expressed in a symbolic form?

About 250 years before the discovery of the Martian moons Kepler (1571-1630) left the following solution of an astronomical anagram of Galileo: *"Salve umbistineum geminatum Martia proles,"* or "Greetings to you, the twin offspring of Mars." Apparently, Kepler was aware of the "twins of Mars"!

Cyrano de Bergerac (1619-1655) in his *Autre Monde* mentioned the two moons of Mars as well. Voltaire (1694-1778) was also certain that Mars had two satellites: "Coasting along the planet Mars, which is well known to be five times smaller than our little earth, they descried two moons subservient to that orb which have escaped the observation of all our astronomers," he wrote in *Micromegas.*

In *Gulliver's Travels,* written in 1726, Jonathan Swift describes the flying island of Laputa, held and propelled in space by a magnet. The scientists on this weightless "space platform" speak about the two moons of Mars. One of these "lesser stars or satellites," as Swift calls them, orbits Mars at a distance of three Martian diameters from the planet's center. The other one whirls around it at a distance of five diameters of Mars.

While the actual distances to the orbits of Deimos, the outer satellite, and Phobos, the inner one, are less than 3½ diameters for Deimos and under 1½ for Phobos, it is true, as Dr. I. M. Levitt, the American astronomer, remarked, that "The similarity between the hypothetical satellites and the real ones was so close that it remains one of the most amazing feats of speculation."*

The previsions of Swift, Voltaire, Bergerac, and Kepler are usually explained in this way—the earth has one moon, Jupiter four (known at the time), therefore Mars must have two. Irrespective of their reasons for believing that Mars has two moons, the writers of the seventeenth and eighteenth centuries certainly hit the target—the planet does have two satellites. What is more, they are the smallest in the whole solar system with its thirty-one satellites. Furthermore, Phobos is the fastest moon in the solar system, as it spins around Mars in seven hours thirty-nine minutes—faster than Mars itself revolves on its own axis. This phenomenon is without parallel in our solar system.

One cannot help thinking that perhaps the ancient Greeks inherited a tradition about these moons of Mars from an unknown source of primordial science. The scientific truth was veiled in their legend of the god Mars with his two companions—Phobos and Deimos.

*I. M. Levitt, *A Space Traveller's Guide to Mars* (Victor Gollancz, London, 1957).

The story of the two moons of Mars, about which people had spoken for 200 years before they actually saw them, is captivating, indeed. However, an even crazier mystery is one which is the exact opposite of the case of the Martian moons. Incredible but true—once upon a time there was a moon, which was seen first, and talked about afterwards. We refer to the strange case of the moon of Venus.

Early in the morning on January 25, 1672, the great astronomer Cassini, who had discovered Jupiter's Red Spot, sighted a smaller object near Venus. He watched it for ten minutes but decided not to create a sensation by claiming the discovery of a Venusian moon. At 4:15 a.m. on August 18, 1686, he saw it again. The satellite was large—one-quarter of Venus in size—situated at a distance of three-fifths of the planet's diameter. The Venusian moon showed phases like the mother planet. Cassini studied the body of fifteen minutes and left complete notes.

On October 23, 1740, James Short of England found "a body" near Venus, one-third of the diameter of the planet, and examined it through his telescope for one hour.

On May 20, 1759, Andreas Mayer of Greifswald, Germany, observed for half an hour an astronomical body in proximity to Venus.

In 1761 Jacques Montaigne, a member of the Limoges Society, who had discovered a comet and had been very skeptical of the Venusian moon observations, saw it himself on March 3, 4, 7, and 11.

On February 10, 11, and 12, 1761, Joseph-Louis Lagrange of Marseille, who later became director of the Berlin Academy of Sciences, reported his sightings of the satellite of Venus.

On March 15, 28, and 29 of the same year Montbarron of Auxerre, France, spotted the Venusian baby-planet through his telescope. Roedkioer of Copenhagen made eight observations of the body in June, July, and August, 1761. The labors of these astronomers finally received a touch of official recognition when Frederick the Great, King of Prussia, proposed that the moon of Venus be named "D'Alembert" in honor of the French savant. Then Christian Horrebow of Copenhagen studied the Venusian satellite on January 3, 1768. What happened afterwards was more mysterious than any unsolved kidnapping that the FBI has ever handled—the Venus baby vanished for a whole century!

It turned up again in 1886, when the astronomer Houzeau saw the mini-Aphrodite seven times. He even baptized it "Neith" in honor of the Egyptian goddess of learning.

On August 13, 1892, the American astronomer Edward Emerson Barnard sighted a seventh-magnitude object near Venus in spite of the fact that he had had no faith in the Venusian moon story. His report is highly reliable because Professor Barnard was the discoverer of the fifth moon of Jupiter and also a star in the constellation Ophiuchus, named in his honor. And while the fifth Jovian moon is still merrily going around its mother planet and Barnard's star has not stopped twinkling, the offspring of Venus had disappeared again.

For 100 years astronomers were on the lookout for this illegitimate baby of the goddess of Love but without any success. The riddle of the Venusian moon, seen by so many astronomers, is still unsolved.

Can the right planet be discovered by wrong calculations or the wrong planet discovered by right calculations? Evidently it can. Forty years ago, after a lot of figuring Dr. Clyde Tombaugh decided that because of Neptune's odd movement, there must be another planet beyond it. He pointed his telescope in the right direction and found the planet Pluto in 1930. But now, after all these years, astronomers say that Pluto could not have disturbed Neptune or Uranus because it is much too small. The discovery of the planet was a freak coincidence, they claim. The moral is this—it can sometimes pay to make errors in your mathematics, if indeed errors were made in this case.

And now we come to the biggest scandal in celestial affairs. On March 26, 1859, a Dr. Lescarbault of Orgeres, France, observed a moving astronomical body on the disk of the sun for 1¼ hours. Leverrier, the director of the Paris Observatory, visited Dr. Lescarbault in order to check on his observation, calculations, and background. This he did with great skepticism and little enthusiasm. However, Leverrier was satisfied with the interview and concluded that an intramercurial planet had been discovered by Lescarbault. He computed its mass to be one-seventeenth that of Mercury, its orbit equal to nineteen of our days, and named it Vulcan.

Dr. Lescarbault presented his findings to the Academy in Paris in January, 1860. Immediately, Napoleon III awarded him the coveted Légion d'Honneur. While France was basking in the glory of this astronomical discovery, Vulcan suddenly refused to parade before the telescopes and vanished as unexpectedly as the moon of Venus. But in this case it was not a mere moon but a whole planet gone astray!

However, to make matters worse, in 1878 Professor James Watson of the University of Michigan claimed to have seen two Vulcans instead of one! An amateur astronomer, Lewis Swift, also had a good look at Vulcan from Pikes Peak in Colorado. But Swift was no ordinary stargazer as his work on nebulae had received recognition by astronomers.

It is sheer impertinence for critics to say that all these men of science were hallucinating and that Lescarbault got his Légion d'Honneur for nothing. The observations were undoubtedly genuine but we still do not know what the body was that crossed the disk of the sun in 1859. Was it an asteroid or a giant space platform from another world? And was the Venusian moon a huge space city cruising the galaxy?

12

Maps, Manuscripts, and Marvels

In the *Life of Apollonius of Tyana* by Flavius Philostratus of
Athens (A.D. 175-249) there is an intriguing passage which
points to unsuspected knowledge of geography in antiquity:
"If the land be considered in relation to the entire mass of
water, we can show that the earth is the lesser of the two."

If the ancient Greeks, Cretans, or Phoenicians had not
crossed the Atlantic or the Pacific, how could Philostratus
know that the oceans cover the greater part of the surface of
the planet?

Plato must have been cognizant of the great size of our
globe and of other continents, because he said in the *Phaedo*
that the Mediterranean people occupied "only a small portion
of the earth."

"Besides the world we inhabit, there may be one or more
other worlds peopled by beings different from ourselves,"
wrote Strabo (first century B.C.). He even mentioned that if
the parallel of Athens were extended westward—across the
Atlantic—these other races might live there in the temperate
zone, clearly alluding to North America.

Yet in the days of Columbus nearly everyone believed the earth was flat, and that the *Nina,* the *Pinta,* and the *Santa Maria* would fall over the edge of this plane if they sailed far enough. Small wonder it was so difficult to recruit the crew for this first transatlantic voyage.

From these historical facts it can be seen that the cognizance of ancient peoples in geography was greatly superior to that of the Europeans of the fifteenth century.

Herodotus (V, 49) tells us that Aristagoras, the ruler of Miletus (500 B.C.), possessed a bronze tablet on which lands and seas were engraved. This might have been one of the earliest maps excepting the clay tablets of the Babylonians.

Only if they had explored distant places themselves could the people of former times have described those places with such accuracy. Pytheas of Marseilles, ancient geographer and astronomer (330 B.C.), sailed as far as the Arctic Circle in the Atlantic and gave a scientiffc explanation of the midnight sun.

Did the scholars of antiquity know about America? Seneca (first century A.D.), the tragedian, confirms this supposition by his famous verse in the *Medea*:

> There shall come a time
> When the bands of Ocean
> Shall be loosened,
> And the vast Earth shall be laid open,
> Another Tiphys shall disclose new worlds,
> And lands shall be seen beyond Thule.

New lands "beyond Thule," or Iceland,* could be nothing but Greenland and North America. Tiphys was the pilot of the legendary ship *Argos*. Seneca's lines definitely allude to what was called the New World centuries later.

In the fifth century B.C. Plato wrote in the *Timaeus* about

---

*Claudius Ptolemy (A.D. 140) placed Thule on the 63° parallel, on which Iceland is situated.

the Atlantic Ocean and America: "In those days the Atlantic was navigable from an island situated to the west of the straits which you call the Pillars of Hercules; from it could be reached other islands and from the islands you might pass through to the opposite continent which surrounded the true ocean."

This phrase implies that beyond the Strait of Gibraltar, the Canaries and the Azores, across the Atlantic Ocean, lies a continent which must be the Americas. This is a momentous statement for it suggests that twenty-five centuries ago or earlier, the ancients were somehow aware of the existence of the Americas.

The *Vishnu Purana,* a sacred book of India, contains a significant passage about Pushkar (a continent) with two Varshas (lands) which lie at the foot of Meru (the North Pole). The continent faces Kshira (an ocean of milk), and the two lands are shaped like a bow. Mythological nonsense? Not really. The Brahmin text concerns the continent of America (Pushkar), with its two land divisions, North and South (the two Varshas). America certainly faces the polar ocean (ocean of milk), and the profile of North and South America does resemble a bow as described by the *Vishnu Purana.*

After this interpretation of the passage from the sacred book of India, a question immediately arises—from where could the Brahmins have got information about America and its exact shape from Greenland to Patagonia? Geographical survey implies means of transport and instruments. But the civilization of India did not have oceangoing vessels 1,500 years before Columbus. And so we have another unsolved mystery in the history of science.

An ancient Tibetan book of the Bon sect contains a strange chart. It is a mosaic of squares and rectangles marked with names of unknown countries. As the diagram shows the four cardinal points—the east on top, the west at the bottom, the south on the right, and the north on the left—the Soviet philologist Bronislav Kouznetsov concluded that the chart was a map.* He found a key to it and identified places such as the Persian city of Pasargady (fourth to seventh century B.C.), Alexandria, and Jerusalem; the countries of Bactria, Babylonia, and North Persia; and the Caspian Sea.

The discovery provides proof of the geographical knowledge of the Tibetans and their links with Persia and Egypt centuries ago, which Orientalists have not suspected until now.

The Yale University map of 1440 conclusively proves that the Vikings reached Greenland and Canada four hundred years before the Spanish landed in San Salvador in 1492. Curiously enough, for their navigation the Vikings used sunstones or special crystals which changed color if pointed toward the sun even in cloudy weather.

The Azerbaijan Academy of Sciences (USSR) made a discovery in 1964 that the thirteenth-century scholar Nasireddin Tusi was aware of the existence of America 250 years before Columbus. The astronomer mentioned the land of Eternal Isles in one of his books, giving its geographical coordinates. When these were joined, the contour corresponded to the east coast of South America. Where did Nasireddin Tusi obtain his information about a faraway continent? In the thirteenth century Mediterranean vessels were too small and unreliable to cross the Atlantic from Gibraltar to Brazil.

The sixteenth-century Turkish cartographer Admiral Piri Reis compiled an atlas called *Bahriye* or the *Book of the Seas,* containing 210 well-drawn maps.

The National Museum of Turkey has in its possession two old maps made by Piri Reis, dated 1513 and 1528. The map with the date 1513 shows Brittany, Spain, West Africa, the Atlantic, parts of North America, and a complete outline of the eastern half of South America. At the very bottom of the

Baikal magazine (USSR), No. 3, 1969.

This prehistoric rock painting near Brandberg, South-West
Africa, depicts young amazons whose European faces are painted
with a light tint and the hair shown in red or yellow. Who were
these European girls in a remote part of Africa?

Babylonian priest-astronomers were aware of the phases of Venus, the four moons of Jupiter, and the seven satellites of Saturn, none of which can be seen without a telescope.

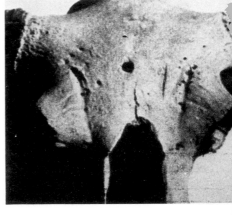

LEFT: A mysterious rock carving near Navai, Uzbekistan, USSR, estimated to be at least 3,000 years old. The men appear to be wearing respirators. Could the object with rays be a space rocket? RIGHT: This auroch's skull is hundreds of thousands of years old. Man, we are told, had neither bows and arrows nor guns then, yet the frontal part of the animal shows a round hole, without radial lines, characteristic of bullet holes.

The Augustine monk Wenzel Seiler and Emperor Leopold I alchemically transmuted the lower portion of this silver medal into gold in 1677. It can still be seen in the Kunsthistorisches Museum at Vienna.

A dragon-chariot in the clouds—an aeronautical idea from ancient China.

Admiral Piri Reis' enigmatic maps dated 1513 and 1528, showing uncharted Antarctica and unexplored rivers of South America. The source data from these maps goes back to the period of Alexander the Great.

ABOVE: Ancient India had notions of aviation as this bas-relief suggests. BELOW: Magnetic Hill, New Brunswick, Canada, where cars go uphill without power, against the law of gravity.

MAGNETIC HILL
TO APPRECIATE THIS
PHENOMENON PROCEED
TO SPOT INDICATED
BY WHITE POST
TURN OFF MOTOR
RELEASE BRAKES

ABOVE: A rock carving near Alice Springs, Australia, depicting a nonaborigine figure in a horizontal position. An archaeological puzzle. BELOW: Ancient cave painting in the Prince Regent River Valley in the Kimberleys in Australia. The figure on the left seems to be wearing a space helmet with an antenna. The presence in Australia of a bearded man in a Babylonian-like miter and of the three European women is a mystery.

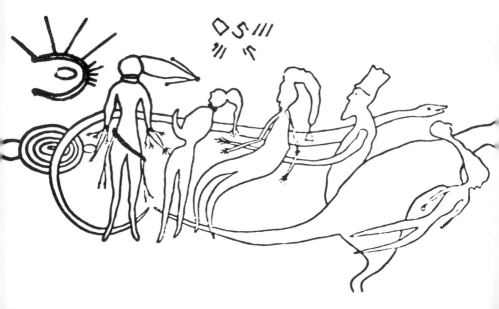

ABOVE: Fragments of an ancient Greek mechanical device found in the sea near the island of Antikythera in 1900 and dated *c.* 65 B.C. In 1959 it was determined that they were parts of a computer —one of the earliest in the world. BELOW: Originally the Antikythera computer, the size of a portable typewriter, looked like this.

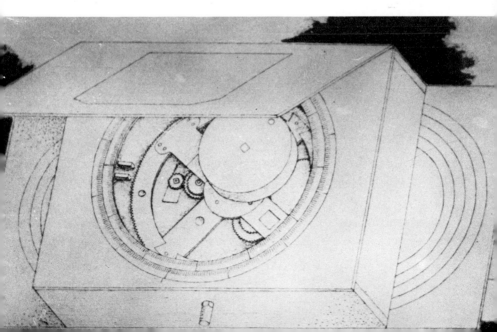

ABOVE: Blocking the Gulf Stream, Atlantis might have been the cause of the last Ice Age in Europe and America. BELOW: The sinking of Atlantis 12,000 years ago (according to Plato) could have ended the Ice Age by letting the warm Gulf Stream advance northward to become "central heating" for Europe.

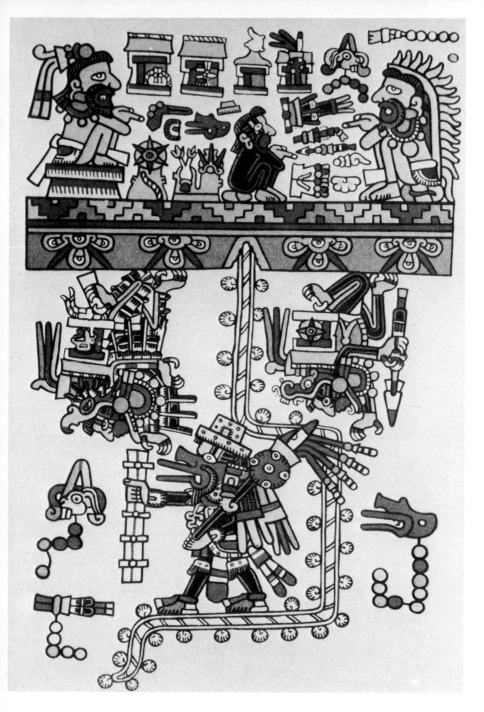

Quetzalcoatl, the Feathered Serpent or the Morning Star, shown descending from the sky on a rope ladder to bring civilization to Mexico. A cosmic culture bearer?

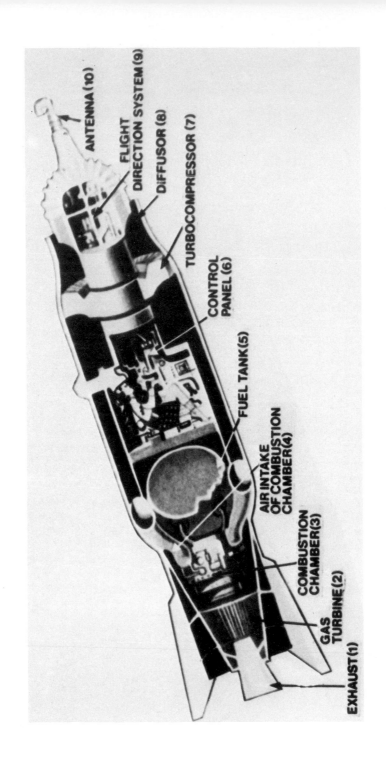

ANTENNA (10)

FLIGHT
DIRECTION SYSTEM (9)

DIFFUSOR (8)

TURBOCOMPRESSOR (7)

CONTROL
PANEL (6)

FUEL TANK (5)

AIR INTAKE
OF COMBUSTION
CHAMBER (4)

COMBUSTION
CHAMBER (3)

GAS
TURBINE (2)

EXHAUST (1)

Soviet science fiction writer Kazantsev
sees a jet rocket in this Mayan carving
on a tomb in Palenque, Mexico, and
identifies all of its mechanical parts.

This heavy boulder in front of a Mohammedan shrine i
Shivapur, India, becomes weightless when touched by eleve

eople chanting "Qamar Ali Dervish." Antigravitation?

The Comte de Saint-Germain —"a man who never dies and who knows everything" according to Voltaire. His riches were attributed to his alchemical skill.

Apollonius of Tyana saw demonstrations of antigravitation and luminous stones in a Trans-Himalayan community.

The Sumerians and Babylonians became civilized when
Oannes, a fishlike monster, gave them the elements of all
the arts and sciences. A cosmonaut in a spacesuit?

Nicholas Roerich in Mongolia with the chest allegedly containing a stone from another planet.

An ancient Greek philosopher? No, the gentleman is a Cro-Magnon caveman from the Crimea who lived some 15,000 years ago.

map is shown the coastline of Antarctica, extending eastward to a point under Africa.

The chart was torn but it is suspected that originally there were three more sections exhibiting the Indian Ocean, and perhaps Australia, Europe, and Asia. This conclusion is suggested by the existing two maps, which appear to be portions of a larger one.

The second map, dated 1528, shows Greenland, Labrador, Newfoundland, a part of Canada, and the east coast of North America to Florida. The geographical projection of these maps could not be determined until recent years. The famous Swedish explorer and savant Nordenskjöld spent seventeen years in trying to solve the projection. His work was completed by the American cartographer Arlington H. Mallery, who had the cooperation of the United States Navy Hydrographic Office.

It was astounding to discover that the maps had been drawn with the greatest precision. The distances between Europe, Africa, and the Americas were exact. Until the eighteenth century navigators could not determine longitude with accuracy. In other words, this sixteenth-century map was superior to later navigational charts.

The text of the atlas *Bahriye* and notations on the Piri Reis' geographical maps by the admiral throw light on the origin of these amazing documents. According to Piri Reis, when he was thirty-one he and his uncle, Captain Kemal, fought against Spain in 1501. In the course of the naval battle they captured a Spanish sailor who had some rare maps on him. The Spaniard told the Turks that he had been on three of Columbus' expeditions and that the discoverer of America had used those maps. If so, the words of Christopher Columbus' biographer Las Casas then become clear: "He was as sure he would discover what he did discover as if he held it in a chamber under lock and key."

Piri Reis himself revealed the story of the maps which he

had secured by questioning the Spanish seaman: "A certain book from the time of Alexander the Great was translated in Europe and after reading it Christopher Columbus went and discovered the Antilles with the vessels he obtained from the Spanish government."*

Fantastic it may sound, but the origin of the maps is apparently traceable to Greece or Alexandria. After studying the two maps Professor Afetinan of Turkey said: "It is quite evident today that Piri Reis came into possession of the map that the great discoverer used."

The two maps, dated 1513 and 1528, pose a number of questions. How could navigators of Piri Reis' time attain such accuracy in cartography? Did ancient Greeks reach and survey South America? Why does the Turkish cartographer's chart of 1513 show not only the coastline of South America but even the unexplored rivers of that continent—the Orinoco, the Amazon, the Parana, the Uruguay, and others? How did Piri Reis learn about ice-free Antarctica?

Now let us review the navigation routes of the New World explorers. The destinations of the three voyages of Columbus made between 1492 and 1498 were the Bahamas, Puerto Rico, and Haiti. In 1501 Vespucci sailed from the coast of Brazil (off Recife) down to Rio de la Plata, where Montevideo is today. Magellan followed his course in 1519, went through the strait bearing his name today, and emerged into the Pacific on his way around the world. Neither Vespucci nor Magellan had explored the rivers of South America beyond the deltas, nor had they carried out any land survey inland.

Yet the 1513 map shows the entire profile of Brazil, which Vespucci could not have charted in 1501 since after reaching Argentina he turned into the Atlantic from La Plata. The map was made six years before Magellan's historic voyage showing the then-unknown shores from what is now Montevideo down

*Prof. Dr. Afetinan, *The Oldest Map of America* (Ankara, 1954).

to Patagonia. What explorer could have drawn this coastline on Piri Reis' chart?

Cortés landed in Mexico in 1520, seven years after the completion of Piri Reis' map. Pizarro occupied Peru in 1531, or eighteen years after the map had been issued.

Antarctica was discovered in the nineteenth century and its charting is still going on. Oddly, the ancient map of Piri Reis shows Antarctica stretching right under Africa, completely free of polar ice, and even indicates the altitudes of mountains which are under glaciers now, and whose height in many cases has not yet been measured! Until the International Geophysical Year probes we knew almost nothing about these mountain ridges sealed under ice.

The riddle of the Piri Reis maps still stands as a challenge to science. Who performed the geographical survey required for drawing these ancient maps with such precision?

As Dr. C. H. Hapgood noted in his book *Earth's Shifting Crust,* "the mapping of Antarctica had actually been done when the land was ice-free." If this is true, the maps of Piri Reis must be copies of charts thousands of years old. Arlington Mallery, the U.S. expert on cartography, adds a touch of mystery to the subject when he says: "We don't know how they could map it so accurately without an aeroplane."

There is no doubt that the Turkish admiral had used some very ancient sources, and this he admitted in a note on one of the maps: "In preparing this map I made use of about twenty old charts and eight Mappa Mundis, i.e. of the charts called *Jaferiye* by the Arabs and prepared at the time of Alexander the Great, and in which the whole inhabited world was shown." This is a clear indication of the antiquity of the data from which Piri Reis drew his charts. Yet another map, that of Orontus Finaeus, dated 1531, can be placed in the same category as those of the Turkish cartographer. The outline of Antarctica is also displayed on this old map. It shows rivers, which implies that the South Pole was warmer in former times than now. Mountain ranges are likewise indicated on the map. They are now covered by a thick ice cap. This document is another enigma because the exploration of Antarctica did not begin before the first part of the nineteenth century.

After having checked the Orontus Finaeus map, Captain Burroughs, Chief of the United States Air Force Cartographic Section, made the following statement in 1961: "It is our opinion that the accuracy of the cartographic features shown in the Orontus Finaeus map suggests, beyond a doubt, that it was also compiled from accurate source maps of Antarctica."

The map of Zeno, with the earlier date of 1380, is also a mystery as it shows Greenland without the ice sheet. The rivers and mountains drawn on this chart have been located in the probes of the French Polar Expedition of Paul-Emile Victor in 1947-1949. This discovery conclusively proves that the source of Zeno's map was very ancient, and that the mapping of Greenland had been done in a temperate climate.

Many conclusions can be drawn from these enigmatic maps. There must have existed an unknown civilization which had oceangoing ships and scientists with a good knowledge of astronomy, navigation, and mathematics in order to have charted Antarctica and Greenland. The vessels employed in these ancient polar explorations had to be large and strong and immensely superior to the craft possessed by ancient Egypt, Phoenicia, Greece, or Rome.

The view expressed by Professor C. H. Hapgood is logical:

The evidence presented by the ancient maps appears to suggest the existence in remote times, before the rise of any of the known cultures, of a true civilization, of a comparatively advanced sort, which either was localized in one area but had worldwide commerce, or was, in a real sense a worldwide culture.

He believes that while palaeolithic peoples lived in Europe, a more advanced culture existed elsewhere. After all, stone-age tribes of New Guinea and Central Australia are with us today in this era of technology. The same situation could have prevailed in the distant past.

Comparable with the mysterious maps of Piri Reis, Orontus Finaeus, and Zeno is the so-called Voynich Manuscript. In 1912 Wilford Voynich, a New York collector of antiquities, found this document inside a locked chest in an ancient castle near Rome. In 1665 the manuscript was in possession of the Jesuit scholar Athanasius Kircher, who received it from a friend with an accompanying letter which said: "Such sphinxes as these obey no one but their master."

The manuscript is certainly an enigma. Although Professor W. Romaine Newbold of the University of Pennsylvania made a serious attempt to decipher the coded document, some of his conclusions have been rejected. Wartime experts who cracked the complex codes of Germany and Japan could not do much with it. An RCA 301 computer was given the problem of interpreting the text and the numerous diagrams of this handwritten book but was not able to clarify the mystery.

The Voynich Manuscript contains well over 250 pages about the format of the book you are reading. On most of the pages there are diagrams in color with captions. There are also 33 pages of text. In the opinion of Professor Newbold, the parchment, the ink, and the style of the drawings indicate the thirteenth century as the time of origin. Other experts think it was written around 1500.

The document is devoted to botanical, astronomical, biological, and pharmaceutical subjects. There are charts depicting cross sections of leaves and roots which could have been observed only with a microscope, but the microscope was not invented until the seventeenth century. One illustration shows a spiral with eight legs, a cloudy mass with stars in the center, and some writing in it. The legend, deciphered by Newbold, reads that the object is within a triangle formed "by the navel of Pegasus, the girdle of Andromeda and the head of Cassiopea." This chart, therefore, may refer to the Andromeda Galaxy, which is invisible as a spiral without a strong telescope.

In studying this chart during the twenties Professor Eric Doolittle of the University of Pennsylvania made the remark that "in my opinion it unquestionably represented a nebula and that the man who drew it must have had a telescope." But if he did not have a telescope, how could the author have observed the Andromeda Galaxy long before the invention of his instrument? And how could he have studied cross sections of plants without a microscope?

On the other hand, if the man who wrote the Voynich Manuscript did in fact employ a microscope as well as a telescope, then adjustments must be made in the history of science, moving these inventions back three and a half centuries, and perhaps acknowledging Roger Bacon as the true inventor of these instruments. Where did Bacon receive his knowledge concerning the microcosmos and the macrocosmos? If it came from ancient alchemical and Hermetic writings, then the source of his discoveries might have been arcane science, come from time immemorial. In case the reality of this mysterious document—the Voynich Manuscript—is questioned, it should be mentioned that in the year 1962 it was on sale in New York for the substantial sum of $160,000.

13

Electricity in the Remote Past

In 1938-1939 a German archaeologist, Wilhelm König, found near Baghdad a number of earthenware jars with necks covered with asphalt and iron rods encased in copper cylinders. König described his find in *9 Jahre Irak,* published in Austria in 1940. He thought they were electric batteries. Electric batteries from ancient Babylon? The idea was far too fantastic; it needed corroboration.

After World War II Willard Gray of the General Electric Company made a duplicate of the 2,000-year-old battery, filling it with copper sulfate instead of the unknown original electrolyte, which had evaporated. The sister battery of the ancient Babylonian vase-shaped cell was tested and it worked! This is conclusive proof that the Babylonians did indeed use electricity. Inasmuch as a number of electroplated articles had been excavated in the same general area, it was assumed that one of the purposes of the battery was electroplating. As similar jars were found in a magician's hut, it can be surmised that both priests and craftsmen kept the knowledge as a trade

secret. The date of the electroplated materials is 2000 B.C.—
that is, they were 2,000 years older than König's ceramic cells.
It should be noted here that electroplating and galvanization
were introduced only in the first part of the nineteenth
century. Once again it is demonstrated how a certain
technological process used 4,000 years ago was rediscovered in
modern times.

The presence of batteries in ancient Babylon indicates that
certain electrical apparatuses must have been used in
antiquity. Professor Denis Saurat* has found evidence of
electrical devices in ancient Egypt. Perhaps these may explain
those mysterious flashes of light from the eyes of Isis that the
devotees of the cult had seen.

Classical authors have made many statements in their
works testifying to the reality of ever-burning lamps in
antiquity. Unfortunately, there is no way of finding out if these
lamps shone by electric light or some other energy.

Numa Pompilius, the second king of Rome, had a perpetual
light shining in the dome of his temple. Plutarch wrote of a
lamp which burned at the entrance of a temple to Jupiter-
Ammon, and its priests claimed that it had remained alight
for centuries.

Lucian (A.D. 120-180), the Greek satirist, gave a detailed
account of his travels. In Hierapolis, Syria, he saw a shining
jewel in the forehead of the goddess Hera which brilliantly
illuminated the whole temple at night. In the same locale the
temple of Hadad or Jupiter in Baalbek was provided with
another type of lighting—glowing stones.

A beautiful golden lamp in the temple of Minerva which
could burn for a year was described by Pausanias (second
century A.D.). Saint Augustine (A.D. 354-430) left a
description of a wonder lamp in one of his works. It was
located in Egypt in a temple dedicated to Isis, and Saint
Augustine says that neither wind nor water could extinguish

*D. Saurat, *Atlantis and the Giants* (Faber & Faber, London).

it. An ever-burning lamp was found at Antioch during the reign of Justinian of Byzantium (sixth century A.D.). An inscription indicated that it must have been burning for more than five hundred years.

During the early Middle Ages a third-century perpetual lamp was found in England and it had burned for several centuries.

When the sepulcher of Pallas, son of Evander, immortalized by Virgil in his *Aeneid,* was opened near Rome in 1401, the tomb was found to be illuminated by a perpetual lantern which had been alight for more than 2,000 years.

A sarcophagus containing the body of a young woman of patrician stock was found on the Via Appia near Rome in April, 1485. When the dark ointment preserving the body from decomposition had been removed, the girl looked lifelike with her red lips, dark hair, and shapely figure. It was exhibited in Rome and seen by 20,000 people. When the sealed mausoleum was opened, a lighted lamp amazed the men who broke in. It must have been burning for 1,500 years!

In his *Oedipus Aegyptiacus* (Rome, 1652) the Jesuit Kircher refers to lighted lamps found in the subterranean vaults of Memphis.

It is evident from the Babylonian batteries alone that electricity was known to the peoples of the Orient in the remote past.

During his stay in India the author was told about an old document, preserved in the Indian Princes' Library at Ujjain and listed as *Agastya Samhita,* which contains instructions for making electrical batteries:

Place a well-cleaned copper plate in an earthenware vessel. Cover it first by copper sulphate and then by moist sawdust. After that put a mercury-amalgamated-zinc sheet on top of the sawdust to avoid polarization. The contact will produce an energy known by the twin name of Mitra-Varuna. Water will be split by this current into Pranavayu and Udanavayu. A chain of one hundred jars is said to give a very active and effective force.

The *Mitra-Varuna* is now called cathode-anode, and *Pranavayu* and *Udanavayu* are to us oxygen and hydrogen. This document again demonstrates the presence of electricity in the East, long, long ago.

In the temple of Trevandrum, Travancore, the Reverend S. Mateer of the London Protestant Mission saw "a great golden lamp which was lit over one hundred and twenty years ago," in a deep well inside the temple.

Discoveries of ever-burning lamps in the temples of India and the age-old tradition of the magic lamps of the Nagas— the serpent gods and goddesses who live in underground abodes in the Himalayas—raises the possibility of the use of electric light in a forgotten era. On the background of the *Agastya Samhita* text's giving precise directions for constructing electrical batteries, this speculation is not extravagant.

History shows that the priests of India, Sumer, Babylon, and Egypt, as well as their confreres on the other side of the Atlantic—in Mexico and Peru—were custodians of science. It appears likely that in a remote epoch these learned men were forced to withdraw into inaccessible parts of the world to save their accumulated knowledge from the ravages of war or geological upheavals. We still are not certain as to what happened to Crete, Angkor, or Yucatán and why these sophisticated civilizations suddenly came to an end.

If their priests had foresight, they must have anticipated these calamities. In that case, they would have taken their heritage to secret centers as the Russian poet Valery Briusov depicted in verse:

The poets and sages,
Guardians of the Secret Faith,
Hid their Lighted Torches
In deserts, catacombs and caves.

Their archaic science still lives today. In 1966 the author visited Kulu Valley in the Himalayas. In the town of Kulu there is a remarkable old temple on a hill, dedicated to the god Shiva. Its special feature is an eighteen-meter iron mast erected near the temple. In an electric storm the pole attracts the "blessing of Heaven" in the form of lightning which flashes down the mast and hits a statuette of Shiva at its base. The pieces of shattered Shiva are than pasted together by the priest and used for the next "blessing." The custom has existed since time immemorial, which would mean that the presence of electric conductors in India has been a reality from the most ancient times.

Tibet is also known to have had miraculous lamps that burned for long periods. Father Evariste-Regis Huc (1813-1860), who traveled extensively in Asia in the nineteenth century, left a description of one of the ever-burning lamps which he had seen himself in that country.

In Australia the author learned of a village in the jungle near Mount Wilhelmina, in the western half of New Guinea, or Irian. Cut off from civilization, this village has "a system of artificial illumination equal, if not superior, to the 20th century," as C. S. Downey stated at a conference on street lighting and traffic in Pretoria, South Africa, in 1963.

Traders who penetrated this small hamlet lost amid high mountains said that they "were terrified to see many moons suspended in the air and shining with great brightness all night long."* These artificial moons were huge stone balls

*United Press by Harold Guard (London, 1963).

123

mounted on pillars. After sunset they began to glow with a strange neonlike light, illuminating all the streets.

Ion Idriess is a well-known Australian writer who has lived amongst the Torres Strait islanders. In his *Drums of Mer* he tells of a story about the *booyas* which he received from the old aborigines. A *booya* is a round stone set in a large bamboo socket. There were only three of these stone scepters known in the islands. When a chief pointed the round stone toward the sky, a thunderbolt of greenish-blue light flashed. This "cold light" was so brilliant that the spectators seemed to be enveloped in it. Since Torres Strait washes the shores of New Guinea, one can perceive some connection between these *booyas* and the "moons" of Mount Wilhelmina.

Tales of similar shining stones come to us from the other side of the Pacific—South America. Barco Centenera, a memoirist of the conquistadors, wrote about their discovery of the city of Gran Moxo near the source of the Paraguay River in the Matto Grosso. In a work dated 1601 he paints the picture of this island city and says: "On the summit of a 7¾ meter pillar was a great moon which illuminated all the lake, dispelling darkness."

Fifty years ago Colonel P. H. Fawcett was told by the natives of the Matto Grosso that mysterious cold lights had been seen by them in the lost cities of the jungle. Writing to the British author Lewis Spence, he said: "These people have a source of illumination which is strange to us—in fact, they are a remnant of civilization which has gone and which has retained old knowledge." Fawcett was in search of ruins of that vanished civilization, and he made claim to having seen a lost city in the jungle. We can believe in Colonel Fawcett's sincerity because he sacrificed his life in that expedition.

The notion that electricity is a new discovery seems to have been punctured by these historical and current reports about perpetual lamps. But the ancients may have used other forms of energy for lighting as well.

14

Did the Ancients Master Gravitation?

The most powerful known force is the energy locked within the atom. Electromagnetic force is one hundred times weaker than nuclear energy. But gravitational force is quadrillions of times weaker than electricity or magnetism.*

Paradoxically, the weakest force is the most difficult to master because we know so little about it. Gravity occupies an exclusive place in physics. What is extraordinary, the discoveries made in the world of the atom have not clarified the mystery of gravitation but, on the contrary, have introduced even more problems within the major problem.

If we could only insulate things against gravitation, they would become weightless. But so far, this has been a fruitless task. Life would be completely transformed if gravitation were conquered. Cars, trains, ships, planes, and fuel rockets, thus being superfluous, would be displayed in museums. Green grass would grow on roads and highways. Houses would float

*Gravitational energy is 10^{36} weaker than electricity and 10^{38} 38 weaker than nuclear power.

in the air and men fly like birds. However, these crazy days seem to be far away because, although antigravitation research is being conducted by some nations, the mystery of gravity has not been solved.

One principal aspect we do know about gravitation is that it varies with mass. The dim companion of the bright star Sirius is composed of matter in so concentrated a state that a cupful of it would weigh twelve tons. Actually, that load is as light as a feather compared with a cupful of the substance of a certain small star in Cassiopea which would show over five million tons on hypothetical scales.

Speculations about the nature of gravity and the possibility of conquering it are not idle. They are of paramount importance in astronautics and aviation because billions would be saved if antigravitation were discovered.

Some of the most incredible tales of antiquity concern levitation or the power to neutralize gravity. François Lenormant writes in *Chaldean Magic* that by means of sounds the priests of ancient Babylon were able to raise into the air heavy rocks which a thousand men could not have lifted.

Is this how Baalbek was erected? The gigantic slab left in the quarry at the foot of the Baalbek Terrace by the Titans who had built it is 21 meters long, 4.8 wide, and 4.2 deep. Forty thousand workers would be needed in order to move this huge mass. The question is, how could such a multitude have had access to the slab in order to lift it? Moreover, even in this brilliant era of technology there is not a crane in the world today that could raise this monolith from the quarry!

Certain Arab sources contain curious tales about the manner in which the pyramids of Egypt were erected. According to one, the stones were wrapped in papyrus and then struck with a rod by a priest. Thus they became completely weightless and moved through the air for about 50 meters. Then the hierophant repeated the procedure until the stone reached the pyramid and was put in place. This would explain the absence of chips on edges of the stone blocks for which the author searched in vain and the joints into which it is impossible to insert a sheet of paper. Even though the Khufu pyramid is no longer the tallest edifice in the world, it is still the biggest megalithic structure on earth.

Babylonian tablets affirm that sound could lift stones. The Bible speaks of Jericho and what sound waves did to its walls. Coptic writings relate the process by which blocks for the pyramids were elevated by the sound of chanting. However, at the present level of our knowledge we can establish no connection between sound and weightlessness.

Lucian (second century A.D.) testifies to the reality of antigravity feats in ancient history. Speaking about the god Apollo in a temple in Hierapolis, Syria, Lucian relates a wonder which he witnessed himself: "Apollo left the priests on the floor and was born aloft."

The biography of Liu An in the *Shen Hsien Chuan* (fourth century) contains an anecdotal case of levitation. When Liu An swallowed his Taoist elixir, he became airborne. But he had left the container in the courtyard and it was not long before the dogs and poultry licked and drank whatever was left in the vessel. As the historical record says: "They too sailed up to heaven; thus cocks were heard crowing in the sky, and the barking of dogs resounded amidst the clouds." What a pity that they did not have sound motion-pictures in those days! However, let us not discard this historical record of China as a mere anecdote because many a tale of the East has become scientific reality. Do we not travel on magic carpets and watch scenes in magic mirrors today?

A Buddhist jataka narrative speaks of a magic gem capable of raising a man into the air if he holds it in his mouth.

The phenomenon of weightlessness is not miraculous to us anymore, because we are now accustomed to seeing our astronauts experience this in space. There must be a scientific explanation for the fact that some people or objects have been shielded from the force of terrestrial gravity in the past.

In antiquity a present-day scientist would have been called a magician. Contrariwise, a magician of yesterday was often a scientist. Simon the Magus, a first-century-A.D. Gnostic philosopher, was a man of this type. This Hebrew thinker said that "fire was the primeval cause of the manifested world and it was dual in character." This is a simple definition of the atomic structure of matter and its polarity. Simon was able to perform miracles by means of his "magical science." His critical biographers, most of them early Christian fathers, described how fire descended from the sky upon objects previously designated by the magician.

There is a story about Simon addressing thousands in Rome on the subject of his philosophy of gnosis, or knowledge. Tradition says that the "spirits of the air" helped him to raise himself high in the air, for Simon was "a man well versed in magic arts." Although Christian historians were not sure of the source of Simon's powers, the power of levitation was nevertheless attributed to him. The magician was also said to have made statues lose their weight and glide in the air.

Iamblichus, fourth-century-A.D. Neoplatonic philosopher, was also known to have floated in the air to the height of half a meter.

Down through the centuries history testifies to the reality of levitation. In his book on the development of aeronautics Jules Duhem tells of a chronicle by Father Francisco Alvarez, secretary of the Portuguese Embassy in Ethiopia in the early part of the sixteenth century. In 1515 Father Alvarez wrote of a monastery on the mountain of Bidjan. Near the Church of the Epistles a four-foot golden rod hung in the air for many a century. This wonder attracted numerous pilgrims to the monastery, and Father Alvarez was quite certain of the genuine character of the phenomenon which none could explain.

Almost two centuries later Dr. Charles-Jacques Poncet, a

French surgeon residing in Cairo, who traveled widely in Ethiopia, saw the same floating baton on the right side of the church in the years 1698, 1699, and 1700. In his *Lettres* (1717) Dr. Poncet says that he suspected a trick when he saw it, and he asked the abbot of the monastery for permission to check the flying stick from all sides. The monk agreed and the French doctor passed his hand below, above, and around the wand on all sides. He writes: "I was speechless with amazement because I could not see any natural cause for so wonderful a phenomenon."

And in 1863 the French explorer Guillaume Lejean visited the Bidjan monastery and he also saw the golden rod in the air several times.

The Catholic Church lists some two hundred saints who were alleged to have conquered the force of gravity. Any scientist who rejects this testimony just because it comes from a religious source should be consistent and likewise discard all ecclesiastical records on the same ground. This would cover records of all possible kinds including, naturally, the condemnation of Giordano Gruno and the case against Galileo.

Saint Christianna, a Christian missionary in Spain in the third century A.D., is reported by Rufinus* to have performed a feat of antigravitation. The King and Queen of Iberia were having a church built and it happened that one column was so heavy that it could not be put in place. The story goes that the saint came to the building site at midnight and asked for divine help in prayer. Suddenly, the pillar went up into the air and remained hovering until morning. The astonished workers had no trouble in moving the weightless column in the air to the right spot, upon which it regained its weight and was easily installed on the pedestal.

At Mount Cassino in Italy there is a large and heavy stone which was traditionally lifted by Saint Benedict (A.D. 448-548)

History, Book I.

by the neutralization of gravity. The stone was intended for the wall of a monastery being built at the time, and the stonemasons could not budge it. Saint Benedict made the sign of the cross on the block, and while the seven men who could not lift it looked in amazement, he raised it alone without any effort.*

King Ferdinand I was a host to Saint Francis of Paula (1416-1507) in Naples. Through a half-opened door he saw the monk in meditation, floating high above the floor of his room.

Saint Teresa of Avila (1515-1582) used to rise in the air frequently and sometimes at the most inconvenient moments, such as during the visit of an abbess or a bishop to her monastery when she would suddenly rise up to the ceiling.

To help ten men struggling to lift an 11-meter cross Saint Joseph of Copertino (1603-1663) flew for 60 meters, picked up the cross in his arms, and installed it in its place. In 1645 in the presence of the Spanish ambassador to the papal court he raised himself and then floated in the church over the heads of those present to the foot of a religious statue. The ambassador, his wife, and the people in the church were all spellbound with astonishment.

The British in India have given many accounts of yogis sitting in a Buddha-like posture either upon air or on water. They often refrained from reporting these phenomena to the press in England for fear of ridicule.

A comparatively recent account (1951) of a case of levitation in Nepal by E. A. Smythies, advisor to the government of Nepal, concerning his young native servant, is worth quoting: "His head and body were shaking and quivering, his face appeared wet with sweat, and he was making the most extraordinary noises. He seemed to me obviously unconscious of what he was doing or that a circle of rather frightened servants—and myself, were looking at him through the open door at about eight or ten feet distance. This went on for about

*Saint Gregory the Great, *Dialogues,* Book II.

130

ten minutes or a quarter of an hour, when suddenly (with his legs crossed and his hands clasped) he rose about two feet in the air, and after about a second bumped down hard on the floor. This happened again twice, exactly the same except that his hands and legs became separated." The episode was not premeditated and Mr. Smythies was stunned to see this phenomenon of the reversal of gravity's force.*

According to the 2,000-year-old *Surya Siddhanta,* the Siddhas, Adepts of High Science, could become extremely heavy or as light as a feather. This ancient concept of gravity as a variable force rather than a constant is in itself very remarkable, for there was nothing in the physical experience of the ancient Brahmins that we know of to indicate that objects could possibly become heavier or lighter.

In a letter dated July 14, 1871, Lord Lindsay relates his strange experience with D. D. Home: "I was sitting with Mr. Home and Lord Adare and a cousin of his. During the sitting Mr. Home went into a trance, and in that state was carried out of the window. We saw Home floating in the air outside our window. He remained in that position for a few minutes, and then glided into the room, feet foremost, then he sat down." The window Lord Lindsay refers to was 23 meters from the ground![†]

The noted British physicist Sir William Crookes also watched performances of levitation by Home. "On three separate occasions have I seen him raised completely from the floor of the room. Once sitting in an armchair, once kneeling on his chair, and once standing up," he wrote in 1874.

Do demonstrations of antigravitation occur today—in this Space Age when it is needed most? Surprisingly enough, the answer is yes.

Shivapur is 24 kilometers south of Poona, Western India. This small, unknown village may hold the key to what

*R. C. Johnson, *Nurslings of Immortality* (Hodder & Stoughton, London, 1957).

†The incident took place at 5 Buckingham Gate, London.

scientists the world over are looking for. A Mohammedan mosque dedicated to the Suff saint Qamar Ali Dervish stands in Shivapur. In front of a one-story building with a neatly painted façade of one door and two windows is a green lawn on which lies a big granite boulder weighing about 55 kilograms.

Often a bearded Muslim priest sits on the steps of the shrine or on the lawn, reading the Koran. When a reasonable number of people—Indians from Bombay or other cities, Mohammedan pilgrims, or even a rare tourist from abroad—assemble in front of the mosque, the priest closes his Koran and greets the visitors.

Then the motley crowd of newcomers—Brahmins, Parsis, Mohammedans, Communists from Kerala, or an occasional foreigner—gather around the boulder on the lawn, eleven in number—no more and no less. Then the priest, or more frequently an attendant, explains to the eleven people standing around the block of granite that they are to lean over and touch the stone with their index fingers, chanting "Qamar Ali Dervish" in loud, ringing tones. As soon as these instructions are carried out, an incredible thing happens—the boulder springs to life and rises into the air to a height of almost 2 meters! The granite block stays in the air for a second and then comes down with a thud so that the participants have to mind their feet.

There is another lighter stone weighing about 41 kilograms which requires the index fingers of nine persons to become weightless.

If there are more than eleven individuals around the bigger boulder, the stone does not budge. If there are fewer—nothing happens either. Also if the words "Qamar Ali Dervish" are not chanted distinctly at a certain pitch, the stone remains on the ground. In this levitation there are three factors—the index fingers of those who are taking part in the phenomenon, their exact number (eleven or nine), and the correct chanting of the name of the Muslim saint. Should any of these requirements not be fulfilled, the manifestation of antigravitation does not take place. This demonstration frequently takes place six times a day and is performed every day of the year.

The interpretation of this demonstration might lead to a very considerable flow of ink. It would be hotly debated by many men of science, some of whom often condemn before examination. The feat has to be seen to be believed. It is plain enough that since thousands of Brahmins, Buddhists, or Parsis take part in this demonstration of weightlessness, Islam has nothing to do with it. There is not the slightest possibility of mass suggestion. (Actually, if one is careless, the stone can crush one's toes when it comes down.) Evidently, everything hinges on sound waves and biocurrents from the fingers. What science should do is study this phenomenon to find out what causes it.

The American writer and socialist Upton Sinclair could hardly be accused of gullibility, and yet he witnessed levitation in his own home in the twenties. In the presence of his friends—a group of scientists and writers—a man with a strange power raised a 15-kilogram table 2½ meters above Sinclair's head! "I was unwilling not to publish it," he said, realizing the importance of the neutralization of gravity to science.

The famous explorer Madame Alexandra David-Neel, who died in 1969 at the age of a hundred and one, wrote about her strange experiences with levitation in Tibet, where she had lived for fourteen years. In her book *With Mystics and Magicians in Tibet* she said: "Setting aside exaggeration, I am convinced from my limited experiences and what I have heard from trustworthy lamas, that one reaches a condition in which one does not feel the weight of one's body."

Actually, the French explorer was fortunate enough to see a sleepwalking lama, or *lung-gom-pa*. These lamas become almost weightless and glide in the air after a long period of training. The lama she saw on her journey in north Tibet leaped with "the elasticity of a ball and rebounded each time his feet touched the ground." Reading these words, one is reminded of Armstrong's "kangaroo walk" on the moon!

The Tibetans warned Madame David-Neel not to stop or accost the lama as that could have caused his death from shock. As this lama passed by with extraordinary rapidity on his undulating run, the French explorer and her companions decided to follow him on horseback. In spite of their superior means of transport, they could not catch up with the sleepwalking lama! In this trancelike state the *lung-gom-pa* is said to be quite aware of the terrain and obstacles on the way, just as a somnabulist who gets out of a window and walks on roofs.

Madame David-Neel was given some very significant information about this levitation. Morning, evening, and night were said to be more favorable for these sleepwalking marches than noon or afternoon. Therefore, there might be some correlation between the position of the sun and gravity.

The power is developed by deep rhythmic breathing and mental concentration. After long years of practice, the feet of the lama no longer touch the earth and he becomes airborne, gliding with great swiftness, writes David-Neel. Amusing as it sounds, she adds that some lamas create artificial gravity by wearing heavy chains in order not to float away into space!

The works of this Orientalist should be studied to understand this Asiatic approach to the task of conquering gravitation, which they have tackled for many a century.

Anomalies in the moon's gravity and its gravitational belts discovered by Apollo 8 caused the craft to be diverted from its course and altitude by these belts or mascons. On our planet gravitational anomalies are not uncommon. One of the most spectacular examples is Magnetic Hill near Moncton, New Brunswick, Canada, where cars travel uphill without power.

Gravity reverses its rule at the bottom of the hill and as drivers switch off ignition and release brakes, their cars are pulled up to the top of the hill by an invisible force. It is generally thought that a pocket of magnetic iron, deep underground, is responsible for this gravity phenomenon. But a magnet or loadestone supposedly attracts metal only.

However, the magnet of Magnetic Hill affects not only metallic objects such as cars but others as well—a wooden stick or a rubber ball, for example. Under certain conditions even water can flow upward in some spots!

It also affects people and many witnesses testify to its unusual influence.

"There is something here in the ground. You feel it in your bones. It makes you shivery. It almost makes you dizzy," writes one tourist.

Another says: "When I was walking and turned around, it gave me a spin. I felt it in my forehead."

"You'd think there was a giant hand pulling you back," another person describes the sensations on Magnetic Hill.

There is a vast difference between electromagnetism and gravity. The phenomenon on Magnetic Hill may help to solve the mystery of antigravitation.

For thousands of years men have experienced the reverses of gravity force. Perhaps the ancients could give us some clue as to how to acquire antigravitation for use in astronautics and aviation.

15

Prehistoric Aircraft

It was Tsiolkovsky who said that first a dream is born, then it is clad in formulas and blueprints, and thus a fancy is materialized. The reason we travel in jets today is that men made real things out of dreams.

One of the first aeronautical designers in the world was Daedalus. He constructed wings for his son Icarus and himself but in piloting his glider the boy flew too high and fell into the sea which is now called the Icarian Sea. The Wright brothers were more fortunate 4,500 years later because the basis for aviation technology had already been developed before them.

It is erroneous to think that Daedalus belongs to mythology. His colleagues—the engineers of Knossos—constructed water chutes in parabolic curves to conform exactly to the natural flow of water. Only long centuries of science could have produced such streamlining. And the streamline is also an essential part of aerodynamics, which Daedalus might have mastered.

Friar Roger Bacon left a mysterious sentence in one of his

works: "Flying machines as these were of old, and are made even in our days." A statement like that, written in the thirteenth century, is enigmatic, indeed. First of all, Bacon affirmed that engines flying in the air had been a reality in a bygone era, and secondly, that they existed in his day. Both possibilities seem to be farfetched and yet history is replete with legends as well as chronicles of airships in the remote past.

Hermes, or Mercury, wore winged sandals and a winged hat which bore him over land and sea with great speed. "But the tales about Icarus or Mercury are only legends," some may say. This is true, yet very often a myth is a fossil of history and the only record of what actually happened thousands of years ago.

Perhaps more striking are the Chinese annals which relate that Emperor Shun (c. 2258-2208 B.C.) constructed not only a flying apparatus but even made a parachute about the same time as Daedalus built his gliders.

Emperor Cheng Tang (1766 B.C.) ordered Ki-Kung-Shi to design a flying chariot. The primeval aviation constructor completed the assignment and tested the aircraft in flight, reaching the province of Honan. Subsequently, the vessel was destroyed by imperial edict as Cheng Tang was afraid that the secret of its mechanism might fall into the wrong hands. These remarks, made casually, imply that the emperor and his sages must have had blueprints of this skyship. Where did they get their know-how of aeronautics?

The Chinese poet Chu Yuan (third century B.C.) wrote of his flight in a jade chariot at a high altitude over the Gobi Desert toward the snow-capped Kun Lun Mountains in the west. He accurately described how the aircraft was unaffected by the winds and dust of the Gobi, and how he conducted an aerial survey.

In the early part of the fourth century Ko-Hung wrote about a helicopter in China: "Some have made flying cars with wood from the inner part of the jujube tree, using ox leather straps fastened to rotating blades to set the machine in motion."

A stone carving on a grave in the province of Shantung, dated A.D. 147, depicts a dragon chariot flying high above the clouds. Chinese folklore is replete with tales about flying chariots.

The subject of aviation was real at the dawn of history, as we can see from the Sanskrit term *vimana vidya,* or the science of building and piloting airships. Concrete ideas concerning aviation were present in an epoch which is regarded as the infancy of mankind.

The Indian classic *Mahabharata,* one of the oldest books in the world, speaks of "an aerial chariot with the sides of iron and clad with wings." An airplane?

The *Ramayana* describes the *vimana* as a double-deck, circular aircraft with portholes and a dome. It flew with the "speed of the wind" and gave forth a "melodious sound." We cannot say the same thing about our jets. A pilot had to be well trained; otherwise no *vimana* was placed in his hands. The prehistoric craft performed maneuvers which only helicopters can do partially today—that is, stop and remain motionless in the sky. The ancient classic gives an account of how the *vimana* soared above the clouds. From that altitude "the ocean looked like a small pool of water." The aviator was able to see the ocean coast and the deltas of rivers.*

The *vimanas* were kept in *vimana griha,* or hangars. They were propelled by a yellowish-white liquid and employed for warfare, travel, or sport. One is amazed at the wealth of detail in this ancient tale and wonders what stimulated it—a fantasy of the authors or actual memory of a bygone Golden Age of Science.

In ancient India six young men constructed a dirigible airship which could take off, fly, and land. The *Pantachantra* contains the full story of this aviation experiment. The

*C.N. Mehta, *The Flight of Hanuman to Lanka* (Bombay, 1940).

138

prehistoric zeppelin was operated by a complex control system, providing a safe, fast flight and perfect maneuverability. Science fiction in early India or a record of a lost technology?

There are two categories of ancient Sanskrit texts—the factual records known as the *Manusa,* and mythical and religious literature known as the *Daiva.* The *Samara Sutradhara,* which belongs to the factual type of records, treats of air travel from every angle. The book contains 230 amazing stanzas about the construction of flying machines. It deals not only with takeoff, cruising for thousands of kilometers, normal and forced landings, but even with possible collisions of aircraft with birds! If this is the science fiction of antiquity, then it is the best that has ever been written.

Air battles and raids were evidently not uncommon in an unknown chapter of history. The same source mentions chemical and biological warfare. *Samhara* was a missile that crippled and *Moha* was a weapon that produced a state of complete paralysis.

The pyramid texts contain a curious interpretation of the purpose of the pyramids as a "ramp to the sky" so that man "may go up to the sky." There are 5,000-year-old images of Isis extant which portray the goddess with folded wings as a female Icarus. Man's inspiration to fly seems to be as old as the birds which he watched.

Down through the centuries men had not only dreamed of airships but tried to construct them. Long ages separate Daedalus from Leonardo but their ideas were identical—to provide mechanical means whereby man could fly in the air. In these days of fast-flying jets spanning oceans and continents in a few hours, we can only admire the men of the past who had the daring and imagination to detach themselves from the limited thinking of their contemporaries.

At the dawn of our supercentury Professor Simon Newcomb

wrote that there were no materials or machinery whereby "men shall fly long distances through the air" and presented formulas to prove his point. The Wright brothers ignored the equations of their learned compatriot and built the first airplane soon after Newcomb's declaration.

In his youth the famous space expert Professor Hermann Oberth once sat at a table in a restaurant in Lindau on the Bodensee Lake. Pointing at a man at another table, the waiter sarcastically told Oberth: "Look at him! There goes that crazy man who wants to fly." The crazy man's name was Count von Zeppelin.

Folklore throughout the world has strange tales about flying machines. In 1958 the Smithsonian Institution published the results of American, Soviet, and Indian archaeological research which indicated that 10,000 years ago the Eskimos lived in Central Asia. How did they reach Greenland? Possibly on foot or sleighs but Eskimo legend says that they were brought to the arctic north by "giant iron birds." Near Madison, Wisconsin, one can see on the ground from a great distance colossal gravel carvings of birds that measure 62 meters from one wing tip to the other. They look like planes.

A photographic survey by the Peruvian Air Force of the arid tablelands of Nasca showed a network of lines and geometrical figures on the ground as far as the eye can see. The lines were made by human hands, removing darker stones from the soil and exposing the lighter inner layer—an undertaking that must have taken many years to complete.

There are contours of animals and birds besides triangles and trapezoids. Most of the lines run in such a way that there is no connection between them and the more recent Inca roads. On this relief map the pilots can see long-dead streams as well. The age of the Nasca patterns was thus estimated to be at least 1,500 years. The Indians say that the giant pictures on the ground were made by another race before the advent of the Incas. But what is really puzzling is this—these designs and lines can be seen from the air only at an altitude of over 330 meters. For whom were the markings intended—angels, birds, or men in flying machines? It is a notable fact that the area covered by these beacons and signals is of vast proportions, covering hundreds of square kilometers.

In Salvador an antique vase has been found which shows human figures in a dirigible in flight. Is there any connection between this craft and the Nasca lines?

It is not out of place to discuss aviation in the distant past. The progress of science was inspired by tales of flying carpets and chariots in the sky. That is what induced Leonardo da Vinci to begin his work on theoretical aerodynamics.

This century is the noon of science and technology. But there must have been other days. The people of antiquity were the first to see the light of science, but with the Dark Ages came its total eclipse, which was dispersed only a few short centuries ago.

Fabulous as these historical records and legends about aviation are, they fade into insignificance before another notion well established in the days of old relative to interstellar travel and spaceships.

16

They Conquered Space Long Before We Did

Man's desire to fly not only in the air but amidst the stars was present in ancient times. The 4,700-year-old Babylonian *Epic of Etana* contains the poem of the "Flight of Etana":

"I will take you to the throne of Anu," said the eagle. They had soared for an hour and then the eagle said: "Look down, what has become of earth!" Etana looked down and saw that the earth had become like a hill and the sea like a well. And so they flew for another hour, and once again Etana looked down: the earth was now like a grinding stone and the sea like a pot. After the third hour the earth was only a speck of dust, and the sea no longer seen.

Anu, the Zeus of Babylonian Olympus, was the god of the Heavenly Great Depths—which we now call space. The description of this space flight depicts exactly what happens when man leaves the earth.

Whether or not the flight of Etana on the back of an eagle to the throne of Anu (another planet?) was the product of fantasy or a fossil of prehistory in which man had already become familiar with space travel matters little. The point that is really important is the presence of a concept of the round earth which becomes small because of perspective as distance increases.

"This place has no air, its depth is unfathomable and it is black as the blackest night." A description of dark, airless space by an astronaut? No, an extract from the Egyptian *Book of the Dead,* over 3,500 years old!

The Book of Enoch also contains strange things. That may be the reason it was rejected by the bishops and the rabbis as apocryphal. But their wisdom can now be questioned because the Book of Enoch was found incorporated in the Dead Sea Scrolls together with the other books of the Bible. As the Scrolls represent the oldest Bible in the world, going back to the second century B.C. instead of the tenth-century Masoretic Text, the Book of Enoch should be accepted as an authoritative source.

Here are extracts from this obscure scripture:

And they lifted me up into heaven . . . (14:9)
And it was hot as fire and cold as ice . . . (14:13)
I saw the places of the luminaries . . . (17:3)
And I came to a great darkness . . . (17:6)
I saw a deep abyss. (17:11)

Does this not look like a graphic account of space? It is a dark abyss, where objects get hot on the side illuminated by the sun and icy cold on the shaded side. And it is the abode of the sun, moon, planets, and the stars, as Enoch said. He described his emotional reaction at the sight of interstellar space as "fear covered me." To say the least, the words of the prophet are very intriguing, even if this interpretation is not altogether correct. But is there yet another hypothesis?

In the second century of our era Lucian, the Greek author who visited Asia Minor, Syria, and Egypt, wrote his novel *Vera Historia*. He drew a picture of a voyage to the moon which anticipated the trip of Apollo 8: "Having thus continued our course through the sky for the space of seven days and as many nights, on the eighth day we descried a sort of earth in the air, resembling a large, shining circular island, spreading a remarkably brilliant light around it."

Many centuries later people thought that the moon was only a head of cheese. The scholastics would have laughed at the concept of an earthlike moon. What made Lucian picture Luna as an astronomical body which could be reached in eight days? Was it mere fancy or the cosmic science of the wise priests of Egypt and Babylon that inspired him to write about a voyage to the moon?

Chinese historical tradition mentions Hou Yih (or Chih-Chiang Tzu-Yu), the engineer of Emperor Yao who was acquainted with astronautics. In the year 2309 B.C. he decided to go to the moon on a celestial bird. This bird advised him of the exact times of the rising, culmination, and setting of the sun. Was it the equipment of a spaceship which provided this information to the prehistoric astronaut? Hou Yih explored space by "mounting the current of luminous air." The exhaust of a fiery rocket?

Hou Yih flew into space, where "he did not perceive the rotary movement of the sun." This statement is of paramount importance in corroborating the story because it is only in space that man cannot see the diurnal movement of the sun.

On the moon the Chinese astronaut saw the "frozen-looking horizon" and built there the "Palace of Great Cold." His wife Chang Ngo also dabbled in space travel. According to the ancient writings of China, she flew to the moon, which she found a "luminous sphere, shining like glass, of enormous size and very cold; the light of the moon has its birth in the sun," declared Chang Ngo.

It is this message from the moon which makes the 4,300-year-old tale so provoking. Chang Ngo's moon exploration report was correct. Apollo 11 astronauts found the moon desolate with a glasslike soil. It is cold in the shade, colder than at our poles.

The *Collection of Old Tales* compiled in the fourth century from more ancient sources by the literary man of China Wang Chia includes an interesting story from the times of Emperor Yao when Hou Yih and Chang Ngo went to the moon. An enormous ship appeared on the sea at night with brilliant lights which were extinguished during the day. It could also sail to the moon and the stars, hence its name "a ship hanging among the stars" or "the boat to the moon."* Did someone beat us in the space race 4,300 years ago? This giant ship which could travel in the sky or sail the seas was seen for 12 years.

The Chinese poet Chu Yuan (340-278 B.C.) wrote these lines in *Li Sao* in which one can perceive a possible long cosmic trip with the bright sun shining on the black background of space:

> The Driver of the Sun I bade to stay,
> Ere with the setting rays we haste away.
> The way was long, and wrapped in gloom did seem,
> As I urged on to seek my vanished dream.

Perhaps space travel has been a dream for centuries. "The way was long and wrapped in gloom," says the king of Chinese poets, and thus will it seem to astronauts on distant voyages in interplanetary space.

The ancient book of China *Shi Ching* says that when the Divine Emperor saw crime and vice rising in the world, "he

China Reconstructs magazine (Peking, August, 1961).

commanded Chong and Li to cut off communication between the earth and the sky—and since then there has been no more going up or down." Is this not a clear indication of the cessation of space travel in the past?

It is difficult to ascertain, after so many centuries, what method of propulsion was used by the spaceships of antiquity. However, the description of the public baths of Isfahan in ancient Persia by Haman of Sheikh Bahai might unexpectedly help us to unravel this mystery.

This large bathing establishment was supplied with hot water by a crucible of special metal and construction, heated by the flame of a single candle! Was it an unknown alloy or an ingenious mechanism that could amplify the energy of the fire of the candle thousands of times?

The publication *Isfahan,* issued by the Iranian government in Teheran in 1962, apparently realizes the vast importance of this discovery to science: "Of course, somebody has excavated the site of the crucible and taken away the secret from which, undoubtedly in due course, the secret of eternal fuel for moon-bound rockets will be found."*

One of the world's oldest books on astronomy is the Hindu *Surya Siddhanta.* It speaks of *Siddhas* and *Vidyaharas,* or philosophers and scientists, who were able to orbit the earth in a former epoch "below the moon but above the clouds."

Another book from India—the *Samaranagana Sutradhara*—contains a fantastic paragraph about the distant past when men flew in the air in skyships and *heavenly beings* came down from the sky. Was there a sort of two-way space traffic in a forgotten era?

In his essay on the *Rig Veda* Professor H. L. Hariyappa of Mysore University writes that in a distant epoch "gods came to the earth often times," and that it was "the privilege of some men to visit the immortals in heaven." The tradition of

*E. Sykes, *The Extraterrestrials* (Markham House Press, London, 1967).

India is insistent upon the reality of this communication with other worlds during the Golden Age.

Old Sanskrit texts speak of the *Nagas,* or Serpent Gods, who live in underground palaces lighted by luminous gems in the fastness of the Himalayas. The *Nagas* are flying creatures who go on long voyages in the sky. The belief in the *Nagas* is so firmly imprinted in the national consciousness of India that even today motion pictures and stage plays exploit this theme to the delight of Indian audiences. The subterranean city of the *Nagas*—Bhogawati—brilliantly illuminated by diamonds, may perhaps be a folklore image of a space base, lighted and air-conditioned. We wonder if those cosmonauts are still there.

The god Garuda is thought by Brahmins to be a combination of man and bird who travels through space. He is believed to have reached the moon and even the Pole Star, which is fifty light-years away from us.

The fifth volume of the *Mahabharata* contains a passage which has but one meaning—that of life on other planets: "Infinite is the space populated by the perfect ones and gods; there is no limit to their delightful abodes."

The thirteenth-century text *Clear-Knowledge* is a treatise on cosmology of ancient India compiled by the Tibetan scholar Pagba-lama. This old book describes how the first men on earth were created by gods. This primordial race had the ability to fly in the sky as their makers did, but eventually they lost the power to travel in the air.

This legend is supported by another source—the "Royal Pedigrees of Tibetan Kings." The date of this document is the fourteenth century but it can be traced back to the seventh century. There is nothing ambiguous about this historical work, mentioning the reigns of the kings and the main events of 2,000 years ago. The text says that the first seven kings came from the stars. They were able to "walk in the sky," whither they returned. It is obvious that the ancient Tibetans believed in visitors from space.

A rock painting found in 1961 in the mountains of Uzbekistan, USSR, near the town of Navai, depicts a man in a haughty posture standing inside a vehicle surrounded by rays. The men around him seem to be wearing strange respirators on their faces. For lack of any other explanation, it has been suggested that this 3,000-year-old painting might portray the landing of a space rocket and its cosmonauts with breathing devices which they might have needed in our atmosphere in exploring the earth. A goat and a man with a lamb, a kneeling woman without the mask, and a man on his knees wearing a respirator and paying homage to the being in the vehicle are some of the features of this enigmatic picture.

A journey to the source of the Prince Regent River, west of the Kimberleys in Australia, was made by Joseph Bradshaw, who discovered strange paintings in caves about eighty years ago. The figures of men and women drawn were totally unlike the aborigines—in fact, they were of European type.

The women have delicate hands and feet and strange hairdos, and one bearded man wears a miter, or a crown. But there is another slender figure in the painting which is a riddle. It has no eyes, nose, or mouth and its round head looks like a diver's helmet or even a space helmet with a band around the neck. Tassel-shaped adornments are appended to the arms and the waist of the being. A strange oval with two arms protrudes from the helmet.

Behind the group is a snake, which is the aborigines' symbol of "Dreamtime," or the remote past. In the distance a spiral and a horseshoe-shaped object with rays can be seen. Undecipherable hieroglyphics above the picture add to the mystery of this Australian cave painting.

The round-headed being without eyes, nose, or mouth is more than intriguing. One wonders if the oval gadget on the helmet is an antenna and the tassellike appendages on the arms and waist electric wires. Does the spiral represent an orbit and the horseshoe object a spaceship in flight? This painting might be connected with the Australian myth of Guriguda. Where she came from is not known, but Guriguda was covered with quartz crystal instead of skin. According to this myth, she flashed rays of light in every direction. A female astronaut in a plastic space suit?

But the greatest mystery is that connected with the presence of the Europeans on the continent of Australia. Is this something really out of Dreamtime—the past which is so remote that to the aborigines it has become a dream? But what a strange dream for a simple Australian aborigine to have—Europeans looking at a spaceman!

Did the ancients endeavor to communicate with other worlds in the cosmos? This question is not as ludicrous as it may seem. The priests of Babylon, Egypt, India, and Mexico and then the medieval sorcerers practiced an astrological magic in which the spirits or angels of planets and stars were invoked in an elaborate ritual, involving magic circles, signs of the planets and the zodiac, incantation of certain formulas, and burning of specific incenses. It is truly said that magic begins in superstition and ends in science.

These astral worshipers maintained that the twelve zodiacal houses were thrones of celestial hierarchies, or superior intelligences of different levels, which could be contacted by means of these complex rites of stellar magic. In their school of thought they had a concept of a unified solar system of which every planet was an integral part. This belief is now shared by science.

Furthermore, the sages of antiquity had certain ideas concerning cosmic evolution. The Egyptians, Babylonians, Hindus, Greeks, and Chinese all came to the same conclusion—namely, that our planetary evolution is somewhere near the middle of the scale.

The philosophers of the classical age maintained that in

each star there is an intelligence. Achilles Tatius, Diodorus, Chrysippus, Aristotle, Plato, Heraclides of Pontus, Cicero, Theophrastus, Simplicius, Macrobius, Proclus, and Plotinus, as well as the early fathers of the church such as Origen, Saint Augustine, Saint Ambrose, and others held this opinion. Thus in the ancient thought wise men affirmed what modern scientists now say—that the universe is alive.

The ancient cult of the planets and the stars has been so strong throughout the ages that we have a remnant of it in the names of the days of the week. Monday is the day of the moon, Tuesday is the day of Mars, Wednesday is the day of Mercury, Thursday is named in honor of Jupiter, Friday celebrates Venus, Saturday is Saturn-day, and the day of the sun is Sunday.

From the dawn of civilization men have associated planets with gods or angels, having special attributes, who formed a celestial hierarchy. This is as true of ancient Greece and Rome as of Babylon, Egypt, and India.

The sovereign of the gods was Zeus-Jupiter. The begetter was the god of time—Cronus-Saturn—in whose hands was the destiny of the world. In ancient Egypt Saturn was worshiped under the name of Ptah of Memphis. In China the planet Saturn was the Imperial Star.

Now Saturn seems a dim small planet so far as visibility is concerned. It is a mystery why the planet was given such prominence in the pantheons all over the world. How many of our readers have seen it? Because of their brightness and apparent magnitude, Jupiter and Venus should have been the two planets chosen by the ancients for the ruling roles. But the amazing thing is that in size Saturn is the second planet in our system, after Jupiter. How did the people of antiquity know this fact?

Did astronomical knowledge of this nature come from another terrestrial or extraterrestrial civilization? Such a theory, although controversial, could at least pour light on this enigma of mythology.

Tales of the descent of skygods upon earth can be found all over the globe. The New Testament contains a meaningful passage: "Remember to show hospitality. There are some who by so doing, have entertained angels without knowing it." (Heb. 13:2). If angels are ethereal messengers of God, they do not have to be entertained. Yet Lot received his two angelic guests and treated them to a supper. Those angels who ate and drank must have been physical beings, and therefore may possibly have been visitors from another planet.

In Palenque, Mexico, there is a tomb with a strange design in relief. It portrays a conical object with a fiery exhaust. A man leaning forward and holding levers in his hands is inside this rocketlike machine. Archaeologists have given no satisfactory explanation of this enigmatic carving. The Mayan hieroglyphs on the frame of the design on the sarcophagus have been interpreted as the sun, the moon, and the Polar Star, thus giving a cosmic significance to the picture. The dates shown on the tomb are A.D. 603 and 633. In terms of history they are comparatively recent. But the priest buried in the grave was the custodian of the old science of the Mayas, who might have wished to perpetuate the memory of space rockets in protohistory.

Quetzalcoatl, the culture bearer who is still respected by the Mexicans, was called the Plumed Serpent. The *Codex Vindobonensis* portrays him as descending from the sky. A visitor from space?

The Aztecs had a square in Tenochtitlán, now Mexico City, where a significant ceremony was performed. Men with wings were fastened by ropes to a movable cross revolving on a high post. As they threw themselves down the length of the ropes, the cross began to rotate, keeping the bird-men in the air. Often they flew head down as if they were the Sons of the Sun coming down to earth. The place was called Volador (Flying) by the Spanish, and the custom persists to this day. Aside from the fact that this was an allegory of an age-old tradition of the descent of the skygods, it demonstrated man's long-cherished desire to fly.

Myths have developed even in this century. The Cargo Cult of Melanesia is a strange belief that "cargo," or manufactured articles such as knives, tinned goods, soap, or toothbrushes, would be brought to the Stone Age tribes by "big canoes" or "big birds." When American planes dropped loads of foodstuffs in the jungle for the advancing Australian and American troops in 1943, the natives took this as a fulfillment of the myth. After the war they continued to build mock "airstrips" for the big birds to deliver "cargo." They even constructed immense warehouses for the expected goods. Having seen radio installations, they erected masts with aerials and "radio sets" of bamboo by means of which they expected to contact the "gods."

Influenced by Christianity, some thought they could talk to Jesus Christ on these bamboo "radio transmitters." But throughout all of these childish beliefs there were some bases in reality: the "big birds" (airplanes), the "big canoes" (steamships), and the "cargo" (white man's manufactured articles).

In like manner the ancient legends of the "gods descending upon earth" and an era when "men and gods" mixed could be a folk memory of an epoch when skyships of a different technology were seen in our atmosphere.

17

First Robots, Computers, Radio, Television, and Time-Viewing Machines

Automatons, robots, computers, radio, and television existed in former epochs. The presence of devices of this nature in antiquity is as historical as any history can be.

In the second century before our era Egyptian temples had slot machines for holy water. The quantity of water which flowed from the tap was in direct relationship with the weight of the coin thrown into the slot. The temple of Zeus in Athens had a similar automatically controlled holy-water dispenser.

According to Greek legends, Haephestus, the "blacksmith of Olympus," made two golden statues which resembled living young women. They could move of their own accord and hastened to the side of the lame god to aid him as he walked. It cannot be denied that the concept of automation was present in ancient Greece.

Cybernetics is an old science. In China it was known as the art of *Khwai-shuh*, by means of which a statue was brought to life to serve its maker. The description of a mechanical man is contained in the story of Emperor Ta-chouan. The empress

153

found the robot so irresistible that the jealous ruler of the Celestial Empire gave orders to the constructor to break it up in spite of all the admiration that he himself had for the walking robot.

One of the world's first calculating machines was, of course, the 2,600-year-old Chinese abacus. While in China I watched experienced bookkeepers do wonders with this archaic calculator.

But the most perfect computer may be the human mind. In 1959 an amusing incident took place at the University of New South Wales in Sydney when the author lived there. Miss Shakuntal Devi from India, known as the Human Calculator, was asked by the University to face UTECOM, one of Australia's biggest electronic brains.

"Find the cube root of 697,628,098,090," asked the computer men. In seven seconds the Indian girl gave the answer—8,869. The electronic machine produced a slightly different result.

"I'll bet this machine my answer is correct," nonchalantly said Miss Devi. More calculations on the clicking machine and the new result proved that the girl was right after all.

"It's frightening," exclaimed the experts.

The engineers of Alexandria had over one hundred different automatons over 2,000 years ago. The legendary Daedalus, the father of Icarus, is reported to have constructed humanlike figures which moved of their own accord. Plato says that his robots were so active that they had to be prevented from running away! By what energy were they operated?

In the temples of Thebes, Egypt, there were images of gods which could make gestures and speak. It is not at all improbable that some were manipulated by the priests hiding inside them.

According to Garcilaso de la Vega, the Incas had a statue in the valley of Rimac "which spoke and gave answers to questions like the oracle of the Delphic Apollo." Isn't the computer doing the same thing nowadays?

In the old story of Greece of the quest for the golden fleece the Argonauts came to Crete in the course of their voyages and legendary adventures. Medea told them that Talus, the last man left of the ancient bronze race, lived there. Then a metallic creature appeared, threatening to crush the ship *Argo* with rocks, if they drew nearer. A robot? That is what some experts on automation suspect, such as the Austrian scientist Heinz Zemanek.

The know-how of robot construction was recorded in ciphered books on magic and thus preserved for long centuries. The monk Gerbert d'Aurillac (920-1003), professor at the University of Rheims who later became Pope Sylvester II, was reported to have possessed a bronze automaton which answered questions. It was constructed by the pope "under certain stellar and planetary aspects." This early computer said yes or no to questions on important matters concerning politics or religion. Records of this "programming and processing" may still be extant in the Vatican Library. The "magic head" was disposed of after the pope had died.

Albertus Magnus (1206-1280), the Bishop of Regensburg, was a very learned man. He wrote extensively on chemistry, medicine, mathematics, and astronomy. It took him over twenty years to construct his famous android. His biography says that the automaton was composed of "metals and unknown substances chosen according to the stars." The mechanical man walked, spoke, and performed domestic chores. Albertus and his disciple Thomas Aquinas lived together and the android looked after them. The story goes that one day the talkative robot drove Thomas Aquinas mad with his chatter and gossip. Albertus' pupil grabbed a hammer and smashed the machine.

This account should not be dismissed as mere fiction.

Albertus Magnus was a true scholar—in the thirteenth century he explained the Milky Way as a conglomeration of very distant stars. Albertus Magnus and Thomas Aquinas were later canonized by the Catholic Church. The word *android* has even been adopted by science to signify an automaton or robot.

In the civilized world we are convinced that with all the science and technology we have created and the literature we have produced, our languages are superior to those of the so-called primitive races.

But in many American Indian dialects an object has a number of names to denote its changes. A tree or a plant has a different name in each of the four seasons. A tree with blossoms in spring looks entirely different in autumn when it is covered with fruit or red and orange leaves. It has another appearance in summer. While we use adjectives to define these changes in the color and shape of the tree, the American Indian has a separate word for each appearance of the same object. This precision in Red Indian languages cannot be reconciled with the absence of writing in South America.

According to anthropology, when primitive man becomes civilized, he creates an alphabet and literature. The Sumerians and Egyptians invented writing about 4000 B.C. and became culturally mature. From then on, they advanced by leaps and bounds.

But history knows of an advanced civilization in the past which did not have any writing at all—the Incas of Peru. This race produced half of the vegetables we eat today. It built the longest highways in the world, wove the finest textiles, erected megalithic structures, and had a thorough knowledge of astronomy.

When Pizarro conquered Peru he made a startling discovery—Atahualpa and he had one thing in common—neither the conqueror nor the vanquished could read! The conquistador was illiterate, whereas the Lord Inca was unable to read because there was no writing to be read in his empire.

Instead of letters and words the Incas used the *quipu*—that is, cords with knots. By means of colors and knots, they created a mnemonic device accompanied by a verbal comment which was effective enough to record a very complicated system of statistics. Without any figures they had a highly efficient method of accountancy. Without an alphabet, they had literature. There was a superstitious fear of writing which no other race in ancient history had ever shared.

Their official version of the origin of this custom was obscure. At the time of a great epidemic, the state oracle pronounced that men had to do away with writing under the penalty of death. That is how the *quipu* appeared. A mystery shrouds the connection between the plague and the alphabet.

Why did the Incas and their predecessors not have an alphabet or hieroglyphics like all other progressive civilizations? One theory will waken incredulity. If the *quipu* was a remnant of archaic calculators, then the explanation is simple. When the mother technology had disappeared, all that the South American Indians were left with was the equivalent of a stack of punch-hole cards—the *quipu* with no machines.

If there were an advanced civilization with a developed technology which perished in a violent cataclysm, then the presence of this simplified calculator and computer system, known as the *quipu*, could be accounted for as an echo from the past.

There is other evidence pointing to the existence of advanced science in ancient South America. In the Bay of Paracas, Peru, is located the historic Chandelier of the Miraculous Sign of the Three Crosses. This 185-meter-high engraving in rock on a slanted cliff resembles the trident of Neptune with branches. The Spanish savant Beltrán García has promulgated a theory whereby the device, equipped with pulleys and cords, was a pre-Inca seismograph. By means of a pendulum earthquakes could be recorded not only in South

America but anywhere in the world. The huge contrivance is placed exactly in a north-south direction, facing the Pacific. Erosion shows that it is at least 2,000 years old.

The ancients had other apparatuses of this kind, rediscovered centuries later. The Chinese scholar Chang Heng (A.D. 78-139) constructed a seismograph in the form of a vase decorated with dragons holding balls in their mouths. Porcelain frogs were placed around the vessel. During an earth tremor the balls would fall into the mouths of the frogs, depending on the strength of the quake. It was only in 1703 that Jean de Hautefeuille built the first seismograph of modern times.

Chang Heng also invented a globe with metal rings which represented the paths of the sun and other celestial bodies. Powered by a water clock, it was one of the world's first planetariums.

Tao Hung Ching (A.D. 452-536), the Leonardo da Vinci of China, is credited with the invention of a similar astronomical model. It was so fantastically modern that its description should be cited from the original source:

He also built a celestial globe which was about one meter high. The earth was in the middle and remained stationary while the heavens revolved about it. The twenty-eight stellar mansions thus fulfilled their periods, and the Seven Bright Ones (sun, moon, and five planets) pursued their courses. The stars were luminous in the dark and faded in the light. The globe was constantly revolving by a mechanical device, and the whole thing agreed with the actual motion of the heavens.*

The mechanism and lighting effects of this little sixth-century planetarium are most impressive on the background of the Dark Ages, which enshrouded Europe at the time.

But the ancient Greeks were in possession of equally exact

*The Biographies, The Lieh Hsien Ch'uan Chuan.

celestial spheres. According to Cicero (first century B.C.), Marcus Marcellus possessed just such a globe from Syracuse, Sicily, which demonstrated the motion of the sun, moon, and planets. Cicero assures us that the machine was *a very ancient invention,* and that a similar astronomical model was displayed in the Temple of Virtue at Rome. Thales of Miletus (sixth century B.C.) and Archimedes (third century B.C.) were considered to be the constructors of these mechanical devices.

The memory of planetariums has persisted for many a century. The historian Cedrenus writes about Emperor Heraclius of Byzantium, who, upon entry into the city of Bazalum, was shown an immense machine. It represented the night sky with the planets and their orbits. The planetarium had been fabricated for King Chosroes II of Persia (seventh century A.D.).

The National Archaeological Museum of Greece in Athens possesses corroded fragments of a metallic object found by sponge divers near the island of Antikythera in 1900. The complex dials and gears of the mechanism were unlike any artifact from ancient Greece. From the inscription on the instrument and the amphorae found with it, a date *c.* 65 B.C. was ascribed to the object.

The find was registered in the museum as an astrolabe until 1959, when Dr. Derek J. de Solla Price, an English scientist working for the Institute for Advanced Study at Princeton, New Jersey, identified it as the ancestor of our computer.

"It appears that this was, indeed, a computing machine that could work out and exhibit the motions of the sun and moon and probably also the planets," he wrote in *Natural History.* *

The purpose of the gadget was to save tedious astronomical calculation. This discovery was revolutionary because in spite of Greek achievements in scientific speculation, no one suspected them to be so proficient in technology. No wonder

*March, 1962.

Dr. Price wrote in *Scientiffc American** that "it is a bit frightening to know that just before the fall of their great civilization the ancient Greeks had come so close to our age, not only in their thought but also in their scientific technology."

The scientist formed a conclusion that the Antikythera computer had changed all our ideas about the history of science because it certainly could not have been the first or the last of its kind. "Finding a thing like this is like finding a jet plane in the tomb of King Tutenkhamen," said Dr. Price at a meeting in Washington in 1959.

A few years later another momentous discovery was made. After a scrupulous study of Stonehenge, the British astronomer Gerald Hawkins, a professor at Boston University, came to the conclusion that it might have an astronomical significance.

England's Stonehenge, built between 1900 B.C. and 1600 B.C., originally consisted of a 29 ½ -meter ring of 25-ton uprights and horizontal slabs, called the Sarsen Circle, inside of which stood five trilithons or archways. The megaliths were surrounded by a ring of fifty-six holes in the ground, called the Aubrey Circle.

Taking his observations from the center of Stonehenge, Professor Hawkins became convinced that aligning archways, stones, and holes with the sun or the moon, when on the horizon, was one of the purposes of this enigmatic neolithic construction. Returning to America with accurate charts of Stonehenge, Hawkins fed the data into a computer which produced results proving that Stonehenge itself had been a computer!

The professor's findings indicated that the Stonehenge astronomers were so advanced that they had noted a phenomenon undetected even by our modern astronomers— that eclipses of the moon take place in cycles of 56 years. By

*June, 1959.

placing three black and three white stones in certain six holes, and by moving them yearly around the Aubrey Circle, the ancient astronomers of England knew when to expect lunar or solar eclipses. It is of interest to note that the priest-astronomers of Babylon and Egypt were unable to predict eclipses until 1,100 years after the construction of Stonehenge.

Professor Hawkins estimated that the building of this megalithic computer was an effort comparable with the U.S. space program. The energies of one person in a thousand are absorbed in the U.S. space program. Considering the population of England 4,000 years ago, the proportion was the same in the construction of Stonehenge. The question is—what was the source of the scientific knowledge of the people who built Stonehenge?

From prehistoric computers we now come to speaking devices in antiquity. One hundred years ago the Academy of Sciences of France accused Edison and Du Moncel of being charlatans when the gramophone was demonstrated before that learned assembly. The same academic institution would probably smile ironically if the subject of ancient "chanting statues, talking vases" or "speaking stones" were broached.

After so many centuries it is impossible to ascertain *how* they worked but there is no doubt that they were quite effective. In classic times the Romans in Egypt described the singing statue of Memnon erected about 1500 B.C. Musical sounds were heard when the rays of the rising sun illuminated its head. A.D. 130 the Roman Emperor Hadrian listened to this singing monument one morning and heard the sounds three times. Neither did Emperor Septimus Severus (A.D. 193-211) fail to hear the statue "chant" at dawn. After the statue had been subject to certain repairs, the musical sounds stopped. This point proves that the "music" was due to some complicated mechanism triggered by the rays of the sun, which was inadvertently damaged during the restoration work. Tourists can still see the statue of Memnon in Egypt.

The Phoenician Sanchuniathon (*c.* 1193 B.C.?) and Philo Byblos (A.D. 150) spoke of "animated stones." The Christian historian Eusebius (*c.* A.D. 260-340) carried one of these mysterious stones on his chest which answered his questions in a small voice resembling a "low whistling." Arnobius (d. *c.* A.D. 327), another Christian father, confessed that whenever he got hold of a "speaking stone" he was always tempted to put in a question. The answer came in a "clear and sharp small voice." Are we using these stones today calling them transistors?

The Bible mentions teraphim, or images, figures, or heads, which answered questions (Ezek. 21:21 and Gen. 31:34). Maimonides (1135-1204) in *Les Regles des moeurs* says that "the worshippers of the teraphim claimed that as the light of the stars filled the carved statue, it was put *en rapport* with the intelligences of those distant stars and planets who used the statue as an instrument. It is in this manner that the teraphim taught people many useful arts and sciences." Could these have been educational radio programs from galactic civilizations? That is what the wise Maimonides instimates, if we take his words literally.

Seldenus in *De Diis Syriis* alludes to golden teraphim consecrated to a special star or planet and says that they were known to the Egyptians. Other teraphim were mummified heads each with a gold plate under its tongue, on which were engraved "magic words." These Jivaro skulls were mounted on walls and spoke at certain times.

Down through the centuries folklore and ancient writers mention another marvel of antiquity—"magic mirrors." The Book of Enoch says that Azaziel taught men to make magic mirrors, and according to this belief, distant scenes and people could clearly be seen in them. Were they forerunners of television?

Although *Vera Historia* of Lucian is considered to be fiction (in spite of its name), his description of a magic mirror is interesting because there was nothing at the time that could have stimulated his imagination in such a direction: "It was a looking glass of enormous dimensions, lying over a well not very deep. Whoever goes down into this well, hears everything that is said upon earth. And whoever looks in the mirror, sees in it all the cities and nations of the world." Was this science fiction written for idle Roman patricians or was it a chronicle of some past in which magic mirrors were as commonplace as they are today?

Television has successfully replaced magic mirrors, yet a few of these prehistoric devices may have been preserved by those who are still in possession of the secrets of arcane science.

Maxim Gorky, the celebrated Russian writer, related an amazing experience which is all the more convincing because he was a materialist. Early in this century Gorky met a Hindu yogi in the Caucasus who asked the Russian author if he wanted to see something in his album. Maxim Gorky said he wished to see pictures of India. The Hindu put the album on the writer's knees and asked him to turn the pages. These polished copper sheets depicted beautiful cities, temples, and landscapes of India which Maxim Gorky thoroughly enjoyed. When he finished looking at the pictures, Gorky returned the album to the Hindu. The yogi blew on it and smilingly said: "Now will you have another look?"

This is how Gorky himself related the end of the story: "I opened the album and found nothing but blank copper plates without a trace of any pictures! Remarkable people, these Hindus!" he exclaimed.*

Perhaps even more astonishing were the time-viewing devices in antiquity. Franciscus Picus in the *Book of the Six Sciences* outlined the construction of the "Al Mucheff mirror" according to the laws of perspective and under proper astronomical configurations. It is said that in that mirror one

*N. Roerich, *The Indestructible* (Riga, 1936).

could see a panorama of time. If this is true, then the ancients were one step ahead of us—they had time television.

The oracles of Egypt and Greece were famous for predicting future events or restoring scenes of the past. How did they do it?

In his *Readable Relativity* the British scientist Clement V. Durell writes: "But all events, past, present, and future as we call them, are present in our four-dimensional space-time continuum, a universe without past or present, as static as a pile of films which can be formed into a reel for the cinematograph."

The theory of relativity may have been known to the people of ancient times. The *Vision of Isiah* (second to third centuries A.D.) tells a story to that effect. The prophet Isiah was taken to Heaven where he saw God in Eternity. Then the angel, who took him to paradise, said it was time to go back to earth. Isiah was surprised and asked: "Why so soon? I've been here only two hours." But the angel replied: "Not two hours but thirty-two years." The prophet was shocked by these words because he realized that a return to earth would mean old age or death. But the angel comforted him by saying that he would not have aged on his return to earth.

In a photon or antimatter space rocket traveling at a velocity approaching the speed of light, our astronauts would experience an identical shrinkage of time, actually "jumping into the future."

If, for example, all events which take place throughout the universe leave marks on a subatomic level, perhaps the past can be seen. And if, on the other hand, the effects of today's actions are projected into the future, this would mean that the contours of tomorrow exist today. Otherwise, it is difficult to understand how the ancients could have cracked the time barrier and accurately predicted future events.*

The *Chandogya Upanishad* of India says: "Tell me all that

*See *La Barriere du Temps* (Julliard, 1969) by the author.

you know and I will tell you what follows." Would this refer to the programming and processing of the greatest computer—the human mind?

The Oracle of Ammon Ra had an automatic time machine in the shape of a god that could not only walk, talk, and move its head but even accepted scrolls with questions to which he gave intelligent answers. When Alexander the Great confronted it, the image came forward to meet him with the promise: "I give thee to hold all countries under thy feet."

The book of Nostradamus' prophecies, published in 1558, was dedicated to King Henry II. In the introduction, the prophet of Provence forewarns of the persecution of the church and predicts the establishment of a new order. "It shall be in the year 1792 at which time everyone will think it a renovation of the age." The revolution was born of anticlericalism and the Republic of France was founded in 1792. It would seem that Nostradamus had to break the time barrier to prophesy an historical event so accurately.

Like Einstein, Michel Nostradamus writes that "Eternity ties into one—past, present, and future." He admitted that many volumes on magic were studied by him. These ancient books on a lost science, veiled in ciphers and symbols, must have helped Nostradamus to learn the secret of projection in Time.

> An emperor will be born near Italy
> Who will cost the empire dear. (1:60)

Emperor Napoleon I was born near Italy—in Corsica—and he did cost France a great deal—half a million men in the march on Moscow alone.

Michel Nostradamus knew about Hitler, whom he called Hister for two reasons—on account of the Führer's hysterical speeches and his birthplace on the Danube (its Latin name is Ister). He described swastikas as "topsy-turvy crosses" which would roll eastward and overrun Russia (6:49). Doctor Nostradamus saw the Hiroshima atomic bomb in his magic mirror, which he defined as a "great fire" in the land of the "rising sun," or Japan (2:91).

In the first two verses of the *Centuries,* Nostradamus disclosed his method of projecting himself in time. He used a brass tripod, fire, water, and a rod. Although such paraphernalia seem meaningless, an imaginative inventor might one day experiment along this line and perhaps succeed in constructing a time machine.

It is possible that in the vast archives of the Vatican Library, priceless documents exist which prove the reality of arcane science. The manuscripts confiscated by the Inquisition or taken from deceased Catholic scholars must have created a vast pool of recorded knowledge. The instructions on how to make an android, the secrets of alchemy, the mysteries of antigravitation and time travel may all be locked up in the vaults of the Vatican.

These historical accounts about automatons, magic mirrors, talking stones, and archaic computers point to the reality of an extraordinary science in antiquity. No matter how strange, they should be carefully examined to try to find clues concerning new forms of energy or new techniques.

But our century is also a period of revolutionary developments in sociology. In this light it would be interesting to delve into our historic past and uncover the prototype for the "new" economics which were introduced only within the past sixty years.

18

An Enigma of Social Science—the Incas

Science and technology walk in step with the social order of the day. Ctesibius and Heron created a two-cylinder steam engine and a jet engine which worked perfectly. They could have constructed factory machines. But under slavery, labor was so cheap that the steam engine and hydraulic pumps of Ctesibius and the steam turbine of Heron were but curious devices used for opening and closing temple doors or moving the limbs of gods. Social and economic conditions were not yet ripe for the exploitation of these inventions in industry.

While technology cannot flourish under a regime of slavery, speculative philosophy and theoretical science can be developed in a social system in which the scholar does not have to worry about his meals and has plenty of time for contemplation.

The primitive communal society had no private ownership. Hunting was regulated by tribal law and all food was shared equally. This system can still be found in central Australia, in

New Guinea, and on the remote islands of Oceania. This was the sweet infancy of humanity.

In slave-holding societies radical changes in economics took place as private property was established. Masters and slaves, the rich and the poor appeared. Feudalism was an improved version of slavery. And then, with the appearance of cities and free citizens who traded and finally became richer than the nobles, a new class was born—the burghers, or the bourgeoisie. After the downfall of feudalism, the new class of property holders created a new economic system—capitalism.

Socialism is a system of society in which there is no private ownership and all means of production and distribution are in the hands of the state. Ultimately, socialism will abandon money as a means of exchange.

This chapter raises a question of paramount importance— is socialism a new idea? Has it been practiced before? These questions would probably attract furious rejections and abuses from the enthusiastic advocates of Marxism. They would also bring severe criticism from those sociologists who, biased by their tender attachment to capitalism, can see no socialism in the past and hope to see none in the future.

But in history there was an advanced civilization in which all national resources and means of production were owned by the state. It did not even have coins or banknotes, which are still used by socialist countries today. Its populace had the true socialist ideal that work is the purpose of life and not an inevitable evil. That state was the socialist empire of the Incas.

While it is true that the Inca state was ruled by a feudal-type hierarchy of nobles and officials, its economic structure was socialist because feudalism is based on private property, which the Incas did not have. Neither could it be called a primitive communal society, if one considers all the scientific and engineering achievements of the pre-Incas and the Incas.

The moral superiority of the Inca socialist regime can be illustrated by an historical fact reported by Leguisamo in

1589, who had seen 100,000 pesos of gold and silver in the house of an Indian. The door was wide open and only a small stick lay across the threshold to signify that the master of the house was out and nothing was to be touched. Four hundred years later no Peruvian would dream of trusting a little stick to protect him from burglary, and more especially with a fortune in the house.

But the excellence of the Incan state was not only on the moral plane. The Royal Road, 5,230 kilometers long, from Colombia to Chile, was one of the most stupendous engineering projects of man, surpassed only in the twentieth century. The roads of Peru were built so well that cars and trucks roll on them today.

The pre-Incan stonemasonry uncovered at Ollantaytambo and Sacsahuaman is megalithic. The weight of some stone blocks is estimated to be from 20 to 100 tons. Yet in spite of the great mass of the stones, the joints are too tight to permit the penetration of a razor blade. One hundred years ago no contractor in the world could have duplicated such building technique. Only the ancient Egyptians with their 70-ton blocks in the pyramids matched the skill of the South American masons.

In the Cuzco earthquake in 1950, 90 percent of buildings in the city were damaged but not a single Inca-built wall was even cracked! Wherever parts of houses were of Inca origin, it was the Spanish portion of the buildings that toppled or cracked. In a dramatic way the quake proved the superiority of pre-Incan and Incan stonework.

Babylon, Carthage, Greece, and Rome exploited the soil under their feet but gave nothing in return. In this manner they dug their own graves. And are we not heading the same way today in this era of pollution and maximum exploitation of natural resources?

However, the Incan state had a long-range conservation program. Nothing of this sort has ever been done elsewhere at any time in history. None were in want in a land where thousands are starving today. The amazing thing about the Incan and pre-Incan projects was that soil and water resources became richer, not poorer!

It was an empire of paradoxes. There was no money, yet gold was plentiful. The Incas had an excellent postal service across South America but no letters. Since they had no writing, only verbal messages were sent. Although they had no figures or books, their quipu accounting and statistics astonished the conquistadors. Although they had no alphabet or writing, their literature was rich. This was all done by means of the quipu—that is, strings with knots. They had the longest and best roads in the world, but no wheel.

The Incan postal system was the fastest in the world until the middle of last century. Their *chasqui* couriers, running in relays, delivered messages from Quito to Cuzco in five days. The distance is 2,011 kilometers and most of it at an altitude of over 3,000 meters.

In spite of the excellent roads in ancient Rome, it took ten times longer to deliver a letter over a similar distance during the reign of Augustus. With the deterioration of Roman highways after the seventh century, letters began to travel even more slowly.

The Incas had a "telegraph"—a fire-and-smoke code system covering 3,220 kilometers in three hours! Under Incan socialism the telegraph functioned more efficiently than under modern socialism. The author once sent a telegram from a metropolis in a socialist country to a suburban address 25 kilometers away. It was delivered two and a half days later!

Incan socialism was certainly efficient. In spite of a great number of officials, there was no bureausociatic "red tape," for which most present-day socialist states are so notorious. The administration consisted of men completely responsible for the welfare of their region. If a public servant was corrupt or inefficient, and a citizen did not receive the basic necessities of life, the official got a brief order from the Sun Emperor—to jump over a precipice!

In a country where work was a virtue and laziness a crime, all had to toil. No wages were paid but the citizen and his family were supplied with a house, food, and clothing entirely free. Education and medical aid were also free. This system of working for nothing and getting the main necessities of life for nothing is not very likely to be reached by any modern socialist state for some time to come.

When not in the fields, each citizen was employed on some government assignment—road or bridge building, postal service, or army. This so-called *mita* service was so implanted in the psychology of the Indian that the village of Carahuasi continued maintenance work on a bridge for 300 years after the collapse of the Inca Empire. This raises another point— namely, that the economic system of the Incas could not have reached its high level without having had socialism ingrained over a long period.

Our biggest problem today is population control. Yet the Incas were not scared of producing babies. In fact, there were no bachelors or old maids in their empire—those situations were considered to be state offenses. They carried out resettlement programs to avoid a density of populace, yet always showing regard for the comfort of the colonists. For example, the citizen from the Andes was not sent to a tropical valley, or vice versa.

Coins with the seal of Croesus are displayed in our museums. But no one has ever seen a coin of Lord Inca Atahualpa, although he had more gold than Croesus himself. The Inca Empire had a moneyless socialist economy, hence the reason for the absence of coins in South America. Although politically the system had certain feudal features, economywise it was a form of socialism because private property did not exist.

Since all land, natural resources, and manpower were in the

171

hands of the Sun Empire, the Incan state unquestionably practiced a certain form of socialism. As an advanced nation which constructed the longest roads in the world and megalithic masonry that still stands today, and which had many other technical and scientific achievements, it can by no means be classed as a primitive communal society.

The Incan socialism stands out like a red light on the face of history, for no other advanced state had ever managed without private property as the basis of economic relationships. Nor has any other high civilization existed without writing—but the Incas did.

How did it originate—this mystery socialist state in South America? The Incan rulers had fanciful tales about their solar pedigrees and the landing of their celestial ancestors upon this planet. Anthropological discoveries have created a surprising backdrop for the imperial Incas. An analysis of their blood groups has provided stunning evidence of their physiological uniqueness.

Like the Egyptians, the Incas mummified their sovereigns. After death they were enthroned on golden chairs in the Temple of the Sun in Cuzco. Five of these mummies were preserved in the British Museum.* In 1952 Gilbey and Lubran made a blood analysis of the tissues of these mummies with the following results: Mummy Number 3 had blood groups C, E, and c with the absence of D—an almost unparalleled case in the whole world, while Mummy Number 4 had substances D and c with the absence of C and E—an extremely rare combination among the American Indians. Scientists admit that the problem of their origin remains unsolved.*

The Incan monarchs married their sisters in order to secure "an heir to the crown of the pure heaven-born race, uncontaminated by any mixture of earthly mould," as Prescott put it. What was this "heaven-born race" which

*They were destroyed later as water from a broken pipe flooded the storeroom of he museum.

possessed such unusual blood group combinations?

The Inca enigma can be resolved in terms of a remnant of socialism from an unknown civilization. Classical writers such as Virgil in the *Georgics* and Tibullus in the *Elegies* were unanimous in asserting that in ancient times land had been held in common and the doors of houses were always left open as there were no thieves. Reading their works one is reminded of the chronicles about the Incas written by the conquistadors.

The memory of this lost paradise was cherished in antiquity. During the week of the Saturnalia (December 17 to 23) the Roman master served his drinking slaves, took abuse quietly, and tried to please his servitors in every way. This custom is a great puzzle in history. What dynamic tradition compelled the patricians to lower themselves to the level of their slaves?

The holidays of the Saturnalia were celebrated to perpetuate the memory of a Golden Age in which slavery and private property did not exist and all were equal. The freedom allowed to slaves during the Saturnalian week was supposed to be an imitation of the ideal democracy in a past age.

This usage must have been extremely ingrained to be obeyed by the ruling class of ancient Rome. Needless to say, the patricians did not particularly enjoy the hectic week of Saturn in December, during which everything was turned upside down. The noble had to serve food and wine to his dressed-up slaves with a smile as sweet as the one seen nowadays on the faces of experienced waiters in Rome who anticipate a generous tip.

The folk memory of a classless society in a former age persists to this day. The Carnival of Nice before Lent, with its merrymaking, processions, and masked balls, is a direct descendant of the Saturnalia. The same can be said of the *Holi,* the early spring festival of India, in which gay crowds splash colored water, powdered dyes, mud, or tar on each other. A police uniform or the turban of a maharaja are no

Man (bulletin Royal Anthropological Institute, London, 1952).

protection against the dye-splashing revelers as this is the day of the Sudras—the lower class. It is remarkable to see one and the same holiday, celebrated for centuries, in the Mediterranean and in India, geographically so far apart.

We are thus faced with another discovery—like the atomic theory, transmutation, aviation, automation, or electricity—socialism is not new!

If that forgotten Utopia was truly more than a myth, did its leaders create isolated centers to keep alive their culture? The idea that the life of this lost commonwealth continued to be lived in some remote part of the world is present in the writings of Homer, Pindar, Horace, Pliny, Lucian, and many Oriental sources.

The next three chapters are devoted to the men who traveled to one of these oases of secret knowledge and left historical records about the practitioners of the oldest science in the world and marks of their own extraordinary achievements in it.

19

Apollonius Met the Men Who Knew Everything

In the first century A.D. a tall, handsome Greek was asked by a guard at the frontier of Babylon: "What gifts have you brought for the king?"

"All the virtues," replied the Greek.

"Do you suppose our king does not have them?" queried the officer.

"He may have them but he does not know how to use them," answered the bold traveler, whose name was Apollonius of Tyana.

In spite of his provoking manner of speech, the traveler was allowed to cross the Babylonian border as the officials thought that the king himself might be interested in meeting so eccentric a visitor.

Apollonius was born in Cappadocia about 4 B.C. At fourteen his schoolteachers could no longer instruct him because of his inborn intelligence. The boy took the Pythagorean vows at the age of sixteen and attached himself to the temple of Aegae. His wisdom and cures had spread to such an extent that a saying appeared in Cappadocia: "What's the hurry? Rushing to see young Apollonius?"

One day a priest of Daphean Apollo brought him a copper map and told Apollonius that the chart showed the road to the City of the Gods. Soon Apollonius of Tyana was traveling east. In Mespila (Nineveh) a man by the name of Damis offered his services as a guide. The life story of the Greek philosopher was later recorded by Philostratus on the request of the Byzantine Empress Domna.

After a difficult trek from Babylon to India the two travelers turned north from the Ganges in the direction of the Himalayan range. It can be assumed that their destination was Tibet, because the journey took eighteen days.

As the Greek sage was approaching the Asiatic Olympus with his devoted companion, strange things began to take place. The path by which they had come disappeared after them. The countryside shifted its position and they seemed to be in a place preserved by illusion.

On the boundary of this wonderland they were met by a boy who addressed them in Greek as if Apollonius had been expected. Apollonius of Tyana was then presented to the ruler of this land, whom Philostratus calls Iarchas.

This fabulous country was full of scientific marvels. There were wells from which pillars of light projected upward like searchlights. Radiant stones illuminated the town and turned night into day.

Then Apollonius and Damis saw demonstrations of levitation as men became weightless and floated in the air. Four tripods-automatons walked into the dining room to serve food and drinks as the travelers sat down at the host's table. Apollonius' biographer borrows from Homer the description of these robots which "instinct with spirit, rolled from place to place around the blest abodes, self-moved, obedient to the beck of gods."

The technological achievements and intellectual superiority

176

of this community impressed Apollonius so much that he only nodded when King Iarchas stated an obvious fact: "You have come to men who know everything."

According to the philosopher from Tyana, these learned men "were living on earth and at the same time not on it." Does this phrase have an allegorical meaning or literal? If literal, then these people might have had communication with other worlds, particularly since they had mastered gravity. Then we can understand the words of Iarchas that "the universe is a living thing."

Apollonius received a mission from the adepts of Asia. He was to bury certain talismans or magnets in places of future historical meaning. Secondly, he was to shake the tyranny of Rome.

The Greek sage arrived in Rome at the unfavorable period of the persecution of philosophers by Nero and was shortly summoned to a tribunal. As the prosecutor unrolled the scroll with the charges against Apollonius, it was uncannily turned into a blank! Without the charges against him, there was no indictment, and Apollonius of Tyana was set free. From that day on, the Roman authorities began to have a superstitious fear of the wise man of Tyana.

Under Vespasian he fared much better and was appointed counselor to the Roman emperor, and Emperor Titus said to him: "I have indeed taken Jerusalem but you Apollonius have captured me."

During the reign of Domitian he was accused of "un-Roman" activities. At the trial Apollonius looked with disdain at the emperor, whom he had known as a boy. The patricians felt anxiety, remembering the weird things that had gone on at the tribunal of Nero. Domitian and the judges made an attempt to whitewash themselves by withdrawing some of the charges on condition that Apollonius of Tyana would be convicted in the end.

Facing the Emperor of Rome, Apollonius drew his cloak

around him, saying: "You can detain my body but not my soul and, I will add, not even my body." And then he vanished in a flash of light seen by hundreds in the court hall.

History does not mention the date of the Greek sage's death. The centenarian Apollonius is traced to Ephesus and then chroniclers lose sight of this amazing personality.

The stay of Apollonius in Asia, where he studied at the feet of those "who knew everything," is of great historical interest. Apparently, our robots are not new if automatons served Apollonius and Damis in the palace of Iarchas. Antigravitation must have been used by those who could raise themselves and glide in the air. According to the story, the landscape shifted when Apollonius and Damis had arrived at the borders of this secret abode in Tibet. Bending light waves is more of a topic for science fiction than for science but this could explain the wavering scenes on the Tibetan border and the Greek philosopher's disappearance at the tribunal of Domitian. Brilliant light from wells or stones could have been produced by electricity or some other energy.

One has no right to reject the testimony of Philostratus, who used numerous source data at Byzantium, any more than that of Herodotus, Virgil, Plutarch, or any other writers of antiquity. Apollonius was so respected that Septimus Severus, who ruled the Roman Empire from A.D. 193 to 211, had a statue of the Greek sage in his shrine together with those of Jesus Christ and Orpheus.

20

Diamonds and Stars—The Immortal Saint-Germain

There is a curious trait in human character, undoubtedly an atavism from the days of the caveman, to see danger in strangers and the unknown. This trait makes man suspicious and apprehensive of a newcomer who does not conform to the accepted rule of behavior and mode of thinking.

When the Comte de Saint-Germain, the sphinx of the eighteenth century, appeared in England in 1745, it was not surprising that a self-respecting Englishman by the name of Horace Walpole gave him the following characteristics: "He sings, and plays on the violin wonderfully, composes, is mad, and not very sensible."

Certain encyclopedias are even more critical of this mysterious individual and simply call him "an adventurer." But there is an abyss of difference between pinning an epithet on a man and making an objective study of his life and nature. Most of the unfavorable comments about Saint-Germain come from political sources. The French police thought he was a Prussian spy. Other secret services of Europe suspected him of being a Russian agent or a Jacobite of England. However, as Lord Holdernesse wrote to Mitchell, the British Ambassador to Prussia: "his examination has produced nothing very material."

The Enlightened Century gave the world one of its greatest minds—Voltaire, who had a definite opinion about the Comte de Saint-Germain. "He is a man who knows everything," declared the French genius.

One of Saint-Germain's best friends and pupils was Prince Karl von Hesse-Kassel, who wrote *Memoires de Mon Temps,* in which he calls the count "one of the greatest philosophers who ever lived."

The Count Johann Karl Phillip Cobenzl, the Austrian Ambassador to Brussels, also had a very high opinion of Saint-Germain: "He knows everything and shows an uprightness and goodness of soul worthy of admiration."

Reading these remarks about a "man who knows everything," one cannot help thinking about Apollonius and the men in Tibet who also "knew everything."

This probing into the life of Saint-Germain is devoted to his scientific accomplishments and presents evidence showing that, like Apollonius of Tyana, he was a master of that ancient science, the shadow of which we can discern in history and legend.

When the Marshal de Belle-Isle introduced the Comte de Saint-Germain to the Marquise de Pompadour and Louis XV in 1749, the king was suffering from boredom. Madame de Pompadour realized that this stranger could cure the king of France of his ennui. The count had many long conversations with the king and the marquise about alchemy, science, and other subjects.

At first Louis XV was very skeptical of the count's knowledge of chemistry and transmutation. But he could not be too critical of a man whose diamonds were bigger than his own.

The Marshal de Belle-Isle vividly described the first audience of the Comte de Saint-Germain with the king:

"If you can compose the Elixir of Life or find the Philosopher's Stone we will be ready to buy the recipe. Meanwhile, you may have your mansion and your pension," said Louis XV with the final remark: "But I do not say that I believe in your pretensions."

"I require neither a mansion, nor a pension," replied de Saint-Germain sternly. "I bring with me all the means I require, a retinue of servants, money to hire a house," and then he put his hand into his large, braided pocket and took a number of loose diamonds and cast them on the inlaid table that stood between the king and himself in a luxurious hall at Versailles. "And if it please your Majesty, accept these as a poor offering."

His Majesty turned quickly; he could not resist an exclamation of delight when he beheld the stones sparkling with the pure glitter of the spectrum as they lay scattered on the gleaming surface of the polished wood. "There, your Majesty, are some diamonds that I have been able to manufacture by my art."

History does not tell us what facts impressed Louis XV most in their evening discussions at the Trianon, at which no one except the king, Madame de Pompadour, and the count were present.

The magnificent Château de Chambord with 440 rooms was placed at Saint-Germain's disposal, where the idle king discovered the joy of creative labor and began his experiments in chemistry under the foremost chemist of the day—the Comte de Saint-Germain.

As Casanova de Seingalt writes in his famous *Mémoires*:

"The monarch, a martyr to boredom, tried to find a little pleasure or distraction, at all events, by making dyes; according to St. Germain the dyes, discovered by the king, would have a materially beneficial influence on the quality of French fabrics."

The Austrian diplomat Cobenzl watched these experiments of the count and Louis XV and wrote to Kaunitz, saying, "The dyeing of silks was perfected to a degree hitherto unknown; likewise the dyeing of woollens. All this was accomplished without the aid of indigo or cochineal, but with the commonest ingredients and consequently at a very moderate price."

The Countess de Genlis in her memoirs (Paris, 1825) wrote that "He was well acquainted with physics and a very great chemist. My father, who was well qualified to judge, was a great admirer of his abilities in this respect."

There is no doubt that the Comte de Saint-Germain was not only a good chemist but an alchemist as well. The London *Chronicle* for May 31-June 3, 1760, reads:

All we can with justice say, this gentleman is to be considered as an unknown and inoffensive stranger; who has supplies for a large expense, the sources of which are not understood. From Germany he carried into France the reputation of a high and sovereign alchemist, who possessed the secret powder and in consequence the universal medicine. The whisper ran the stranger could make gold. The expense at which he lived seemed to confirm that account.

The count's collection of diamonds and precious stones tended to increase his fame as an alchemist. Baron Charles-Henri de Gleichen, the Danish diplomat in France, wrote his memoirs in *Mercure Étranger,* Paris (1813), about his meetings with Saint-Germain: "But he showed me other wonders—a large quantity of jewels and colored diamonds of extraordinary size and perfection. I thought I beheld the treasures of the Wonderful Lamp."

There were several episodes which supported his ability to transmute metals. When the Marquis de Valbelle visited Saint-Germain in his laboratory, the alchemist asked him for a silver six-franc piece. After covering it with a black substance he exposed the coin to heat of the furnace. A few minutes later the count took it out of the fire. When it cooled the silver piece was no longer of silver but of the purest gold.

Casanova related a similar experiment in his memoirs:

He asked me if I had any money about me. I took several pieces and put them on the table. He got up, and without saying what he was going to do, he took a burning coal and put it on a metal plate and placed a twelve-sols piece with a small black grain on the coal. He then blew it, and in two minutes it seemed on fire. "Wait a moment," said the alchemist, "let it cool." It cooled almost directly. "Take it—it is yours," said he. I took up the piece of money and found it had become gold.

Irrespective of the fact that Casanova could not believe this transmutation, the story is one of a number which are worth examining. The Count Cobenzl witnessed himself "the transmutation of iron into a metal as beautiful as gold, and at least as good for all goldsmith's work," by Saint-Germain.

When a court chaplain at Versailles suspiciously asked Saint-Germain if he dabbled in black magic, the count replied that his laboratory did not deal with the supernatural because he was a serious student of chemistry and had made discoveries that were of some use to humanity.

According to Madame du Hausset, the lady-in-waiting to the Marquise de Pompadour, the count "positively asserted that he knew how to make pearls grow and give them the finest lustre." Cultured pearls were manufactured in China in the thirteenth century. Because of his long stay in the Far East the count could have obtained the secrets of Chinese jewelers and learned how to produce cultured pearls. Saint-Germain was also reputed to have made diamonds. It is incomprehensible that he could have created synthetic diamonds with the limited facilities of his Chambord laboratory. To transform carbonaceous materials into diamonds, a pressure of 273,200 kilograms per square centimeter and a temperature of almost 2,800°C are required.

Madame du Hausset has an interesting entry in her diary:

> There was a discussion between the king, Madame, some lords and the Comte de Saint-Germain about the secret the comte possessed of making flaws in the diamonds disappear. The king sent for a diamond of moderate size which had a flaw in it. It was weighed.
>
> "Its value is estimated at 6,000 livres," the king said to the count, "but it would be worth 10,000 without the flaw. Will you undertake to enable me to make a profit of 4,000?" He examined it carefully: "I may be able to," he said. "I shall return it to your Majesty in a month's time."
>
> A month later the count returned the diamond to the king without a flaw. Louis XV sent it to his jeweler by Monsieur de Gontaut, and he brought back 9,600 livres, but the king reclaimed the diamond to keep it as a curiosity. He never got over his astonishment, and used to say that M. de Saint-Germain must be a millionaire, especially if he had the secret of making big diamonds from little ones. M. de St. Germain did not say either Yes or No.

One of the ministers of France became distrustful of Saint-Germain and decided to check his funds in order to expose the alchemist. As the London *Chronicle* (1760) says:

> He ordered an enquiry to be made whence the remittances he received came, and told those who had applied to him that he would soon show them what quarries they were which yielded this philosopher's stone. But the fact is that in the space of two years, while he was thus watched, he lived as usual, paid for everything in ready money, and yet no remittance came into the kingdom for him.

Frederick the Great gave the following characteristic to the Comte de Saint-Germain: "A man whom no one has been able to understand."

The chemical experiments of Louis XV and Saint-Germain with the dyes—possibly aniline dyes—were conducted long before their actual discovery by Unverdorben in 1826. But what was really incredible—the king supposedly learned how to amalgamate small diamonds into larger ones and thus increase their value manifold.

As Casonova writes in his *Memoires*: " 'I melted down,' said Louis XV, 'small diamonds weighing 24 carats and obtained this large one weighing 12.' The Duc des Deux-Ponts told me this story with his own lips, one evening, when I was supping with him and a Swede, the Comte de Levenhoop, at Metz."

These isolated records draw a picture of Saint-Germain and Louis XV in shirt sleeves amid retorts and furnaces in the Château de Chambord or the Trianon in Versailles.

His fame spread as exciting episodes were told by noblemen after their rare encounters with the count. On one occasion a banquet was served by Saint-Germain in his mansion in Paris and only a select assembly of aristocrats attended it. The dinner had a unique dessert on the menu—a precious stone on a plate, not to eat but to take home!

If the philosopher's stone helped Saint-Germain to

manufacture gold and diamonds, it also served him as an elixir of youth.

The memoirs of people who had known Saint-Germain indicate that he possessed an elixir which on rare occasions he gave to certain persons.

In a letter to Frederick the Great, Voltaire made a significant remark about the Comte de Saint-Germain: "He will probably have the honor of seeing your Majesty in the course of fifty years."

To arrive at a definite conclusion that the count was able to preserve his vigor and youth beyond the limit allotted to man, one has to examine contemporary memoirs, letters, documents, and newspaper articles.

Our first witness is Baron de Gleichen (1735-1807), who said in his memoirs that he had heard "Rameau and an old relative of a French ambassador at Venice testify to having known M. de St. Germain in 1710 when he had the appearance of a man of fifty years of age." Jean-Philippe Rameau (1683-1764) was a famous composer of operas and ballets at the time.

The Marshal de Belle-Isle and Madame du Hausset recorded two scenes which typify the enigma Saint-Germain had created because of his ability to stay young.

MADAME DE POMPADOUR: "Do you say then that you can make the Elixir of Youth?"

ST. GERMAIN: "Ah, madame, all the women want the Elixir of Youth and all the men want the Philosopher's Stone, the one eternal beauty, the other eternal wealth."

P. *"How old are you?"*

ST. G. "Eighty-five years of age, perhaps."

P. "You do not fool me, Monsieur de St. Germain, I shall find out more about your pretensions. I have unmasked quacks and charlatans before."

ST. G. "He who is standing before you, Madame, is your equal. And if I have your permission I shall now take my departure."

The question of the alchemist's age was also brought up in 1758 and Madame du Hausset jotted down the conversation word for word:

MADAME DE POMPADOUR: "But you do not say what your age is, and make yourself out to be very old. The Countess de Gergy who was ambassadoress at Venice about fifty years ago, I think, says that when she knew you there, you were just as you are today."

ST. GERMAIN: "It is true, Madame, that I knew Madame de Gergy a long time ago."

P. "But from what she said, you would be more than a hundred now."

ST. G. "It's not impossible," he said laughingly, "but I think that it is still more possible that the lady, whom I respect, is talking nonsense."

P. "She says you gave her an elixir which was marvelous in its effects, she claims that for a long time she looked only twenty-four. Why should you not give one to the king?"

ST. G. "Ah, Madame, to think of me giving the king an unknown drug! I should be mad," he said with a sort of terror.

Although the count refused to give his elixir to Louis XV, he prepared effective cosmetics for the Marquise de Pompadour which enhanced her beauty to her great delight.

The reminiscences of Rameau and Gergy place our alchemist in Venice about 1710, when he appeared to be about fifty years of age. If so, he was born around 1660, and Saint-Germain should have been a centenarian in 1758, as La Pompadour said.

In the years 1737-1742 he was an honored guest at the court of the Shah of Persia.

In 1745 Horace Walpole, the English author, wrote a letter to Mann in Florence in which he said: "The other day they seized an odd man who goes by the name of Count de St. Germain. He has been here these two years."

In 1745-1746 the count lived in Vienna, where he resided in the mansion of Prince Ferdinand von Lobkowitz.

In 1749 he arrived in Paris on the insistence of the Marshal de Belle-Isle, who introduced the count to Louis XV and Madame de Pompadour.

In 1756 General Robert Clive, the founder of the British rule in India, met Saint-Germain in India.

In 1760 the London *Chronicle* published an article in which the interest that Saint-Germain aroused in London because of his eternal youth was summed up like this: "None now doubted what at first had been treated as a chimera; he was understood to possess with the other grand secret, a remedy for all diseases, and even for the infirmities in which time triumphs over the human fabric."

In 1762 the count stayed in St. Petersburg, where he took part in the *coup d'état* which put Catherine the Great on the throne of Russia. Later in the year, and in 1763, he resided at the Château de Chambord, engaged in his alchemical and chemical experiments.

In 1768 we trace him to Berlin, and in the following year he traveled to Italy, Corsica, and Tunis.

In 1770 he was the guest of Count Orloff when the Russian navy anchored at Livorno, Italy. He wore the uniform of a Russian general. The Orloff brothers had always spoken of the important role the Comte de Saint-Germain played in the Russian "revolution."

In the seventies he stayed in Germany, actively engaged in Masonic and Rosicrucian activities with his protector, friend, and disciple—Prince Karl von Hesse-Kassel.

In 1780 Walsh of London published Saint-Germain's music for the violin, which provides us with another date for the biography of the count.

The church register of Eckernförde, Germany, contains the following entry: "Deceased on February 27, buried on March 2, 1784 the so-called Comte de St. Germain and Weldon. Further information unknown. Privately interred in this church."

The church record does not say when the count was born, nor does it indicate the first name of the "so-called Comte de St. Germain." But if we take the words of the composer Rameau and the Countess de Gergy, he should have been about 124 years old at the time of his death!

One year after his alleged death we find him attending a Masonic conference! *Freimaurer Brüderschaft in Frankreich,* Volume II, page 9, has this notice: "Amongst the Freemasons invited to the great conference at Wilhelmsbad on February 15, 1785 we find St. Germain included with St. Martin and many others."

Stephanie-Felicite, the Comtesse de Genlis (1746-1830), the educationist who wrote over eighty books and received a pension from Napoleon I, made a fantastic statement in her memoirs—she met Saint-Germain in 1821 in Vienna!

The Comte de Chalons, the French ambassador in Venice, also claimed that he had a conversation with the immortal Saint-Germain soon afterwards in the Piazza di San Marco. If we go back to Venice to the year 1710 and recollect the testimony of Madame de Gergy that the count appeared to be fifty years of age at that time, we can calculate that in 1821 he should have been about 161 years old!

The London *Chronicle,* previously referred to, has a story which is typical of the fantastic reports about Saint-Germain's elixir of life that were circulating throughout Europe in the eighteenth century:

A certain duchess feared she should see some of those marks which the crow of age imprints upon the face of beauty. She sent to this stranger: "Monsieur le Count," said she, "what I shall say wants more apology, than I know how to make, but you are all politeness: they tell me you have that inestimable secret, worth more than all your gold, the medicine that will restore youth; I don't know that I want it yet, but time is time; and perhaps, Monsieur, what it can remedy, it will more easily prevent: I would be early in my care; come, answer me, can I obtain it of you? Let me have it, and name your own conditions."

The stranger put on a mysterious air, and answered: "Those who have these secrets do not chose it should be known they have them." "I know it, Sir," replied the Lady, but you may confide in me." In fine, he was prevailed upon: he brought next day a vial of four or five spoonfuls. He told her ten drops was enough to take at once and that only at the new and full of the moon; that it was innocent, but if she wasted this, perhaps, it would not be easy to get a supply.

The Lady put it by in the secure place where she kept her rouge; and went out on a visit. Her woman happened that afternoon to be seized with the colic. She looked over the house for a liqueur (in England we should say a dram) and finding this vial in so careful state of preservation, she doubted not its excellence—she smelt it, 'twas fragrant, she tasted it, 'twas pleasant; and she drank it all off. The colic vanished, and she sat down in great spirits to adjust her Lady's toilet. At evening, the Duchess came in tired, limped to her chamber, and was calling for her things, when casting her eyes upon the woman: "Child," says she, "who are you? What do you want with me? How came you here?" The other only curtsy'd; and the Duchess peevishly bade her go. She answered: "Your

Grace is pleased to speak in an uncommon manner; I have the honor to be your Grace's woman, and wait to undress you." "Heaven and earth," replied the Duchess, "you, my woman! Why child, my woman is five and forty, you, I dare swear, have not yet been sixteen!"

The mystery was never explained, all France rung with the miracle; but the stranger was gone, and the Duchess is now as grey as other matrons of sixty-five, never having been able to obtain another bottle.

This tale can be considered as an anecdote but, on the other hand, the Comte de Saint-Germain's own extreme age and vigor were something that could not be elucidated without the hypothesis of the elixir. Perhaps the great Voltaire was right when he said: "He is a man who never dies."

Franz Graffer recorded in his *Memoirs of Vienna* the significant phrase of Saint-Germain: "I set out tomorrow evening. I shall disappear from Europe and go to the Himalayas." Apollonius of Tyana also went across the Himalayas and found the men "who knew everything." The statement in Graffer's memoirs is important as it gives the location of a center of the sages who have preserved arcane science for thousands of years.

The Comte de Saint-Germain once said that "One needs to have studied in the pyramids as I have studied." So we have on the one hand the secret temples of Egypt, and on the other, the Tibetan monasteries as the source of his knowledge. Viewed from this angle, this poem of Saint-Germain's is hardly an exaggeration:

Curieux scrutateur de la Nature entière,
J'ai connu du grand tout le principe et la fin.
J'ai vu l'or en puissance au fond de sa rivière
J'ai saisi sa matière et surpris son levain.

The only manuscript known to have been written by Saint-Germain is *La très Sainte Trinosophie,* preserved at the Bibliothèque de Troyes.

This document contains symbolic illustrations and enigmatic text. Section 5 has some very strange sentences:

> The velocity with which we sped through space can be compared with naught but itself. In an instant, I had lost sight of the plains below. The earth seemed to me only a vague cloud. I had been lifted to a tremendous height. For quite a long time I rolled through space. I saw globes revolve around me and earths gravitate at my feet.

It does not take a great deal of imagination to recognize in this passage a record of a long space flight in which the earth became small as it did to the Apollo crewmen. But Saint-Germain must have gone farther than the moon because he seems to have reached the planets.

Besides the space barrier, Saint-Germain may have cracked the time barrier as well. "I am much needed in Constantinople, then in England, there to prepare two inventions which you will have in the next century—trains and steamboats," he said to Graffer.

Transmutation, extension of life, space travel, time conquest—all are frontiers of science. It can be surmised that the Comte de Saint-Germain had access to the secret fountain of knowledge.

21

In the Abode of Wisdom—Roerich

Across the Seine from Saint-Germain-des-Prés where the Comte de Saint-Germain lived in Paris, and whose name he probably adopted for his surname, is the Théâtre du Chatelet.

A few generations after his time, in May, 1909, to be exact, the theater staged *Prince Igor* by Borodin and *La Pskovitine* by Rimski-Korsakov. The immortals of the Ballet Russe— Pavlova, Karsavina, and Nijinsky—were in the production. The colorful settings created a sensation and attracted the attention of the Paris press. They were the work of Nicholas Roerich, a young Russian painter.

Since Leonardo da Vinci, Roerich was certainly one of the most inspirational of masters. His rich colors and simple contours and the clear-cut concepts behind the paintings make a powerful impression.

Nicholas Roerich has justly been called the prophet of the canvas as students of his art have listed titles of his paintings from 1897 to 1947 and discovered that many of them were symbolical predictions of coming events.

In 1901 he exhibited his painting *The Ominous* representing black ravens on the background of gray rocks, somber sea, and cloudy sky. It was an anticipation of the horrible wars we have had in this century.

In 1913 Roerich produced his *Human Deeds,* which depicts gray-haired old men gazing at the ruins of a city in wonder and sorrow. *The Doomed City,* showing a walled city encircled by a huge python, was completed early in 1914 and had similar forebodings of a calamity which would destroy cities. *The Three Crowns,* displayed in St. Petersburg about the same time, represents three kings with swords whose crowns are raised from their heads to vanish in the clouds. This work augured the end of the dynasties of the Hohenzollerns, the Hapsburgs, and the Romanovs in World War I, which broke out a few months later.

In March, 1914, Nicholas Roerich exhibited his *Lurid Glare,* which showed a knight standing by a lion in front of a castle, behind which the sky was aflame. As Kaiser Wilhelm's armies swept over Belgium in August of that year, the prophecy was fulfilled. The painting depicted King Albert of Belgium in the fire of war, as Roerich explained its meaning himself.

On the eve of Nazi invasions, Roerich produced several tempera pictures of people with scant belongings trudging on a country road, on the backdrop of a dark sky. It was a prediction of the plight of European refugees in World War II.

"Culture and Peace are the most sacred goals of humanity," wrote Roerich in one of his articles. As if anticipating a world calamity, he founded his pact in Washington, D.C., in 1929 endorsed by the republics of the Pan-American Union. This "Red Cross of Culture" was to protect all cultural centers from destruction by means of a special flag of three red dots in a red circle on a white banner.

If Apollonius of Tyana challenged the might of ancient

Rome and fought for moderation in government, if Saint-Germain played with diamonds in Versailles in order to gain the attention of the ruling class and warn of the impending revolution, then Roerich tried to neutralize the forces of darkness which were pushing mankind into a global war.

Roerich was more than a great master painter, explorer, savant, author, and peace maker. He was an emissary of those who have for ages guarded the fountain of ancient wisdom. The distinguishing marks of these illustrious envoys are possession of secret knowledge which extends beyond the frontiers of our science, humanism, and spirituality.

Science is successfully conquering space but has done little to dispel the mystery of time. Only twenty years ago we scoffed at the idea of a trip to the moon or the planets. Today we smile at the dream of building a time-viewing machine. Yet masters of arcane knowledge have broken the time barrier and correctly predicted future events. The emissaries of the elect philosophers have been equally proffcient on both frontiers of science—time as well as space. In this connection unknown facts from Roerich's life will be disclosed here.

The 1925-28 Roerich expedition set out from Darjeeling, India, across the Himalayas, traversing the cold plateau of Tibet, crossing the mighty Kun Lun Ridge, then emerging into the vast Gobi Desert.

In the caravan there was a pony with a chest tied to its back which Tibetan and Mongol bearers guarded constantly. Roerich's painting *Chintamani,* exhibited in the Roerich Museum in New York, portrays this pony with the coffer in a deep mountain gorge in the Himalayas. The chest on the pony's back is represented with a dazzling radiance of magnificent colors.

There is a mystery surrounding this chest. However, Nicholas Roerich and his wife, Helena, lifted the veil of obscurity in their books, and when one reads between the lines, startling disclosures become evident.

In the *Legend of the Stone* by Helena Roerich the opening words are: "Through the desert I come—I bring the Chalice covered with the shield. Within it is a treasure—the Gift of Orion." This passage alludes to the coffer carried by the pony depicted in Roerich's painting *Chintamani*. What is the "Gift of Orion"? Madame Roerich refers to the *Book of Lun,* which reads: "When the Son of the Sun descended upon earth to teach mankind, there fell from heaven a shield which bore the power of the world." A meteorite or an artifact brought by cosmonauts from another solar system, possibly Sirius, the Dog Star of Orion, the Hunter?

On Eastern Crossroads by Helena Roerich records the appearance of the stone: "The length of my little finger, of greyish luster like a dried fruit, or heart, with four unknown letters." Its radiation is stronger than radium but on a different frequency. This is what the Eastern legend says about the stone. The rays cover a vast area and influence world events. The main mass of the stone is kept on a tower in the City of the Starborn—Shambhala in Central Asia.

Was Roerich the emissary who carried the stone through many countries back to its temple in Shambhala? "The miracle of Orion's rays is guiding the people," asserts the Oriental legend. How did it reach the earth? Was it a meteorite of a unique nature or was it brought to earth in a spaceship by our elder brothers in evolution from another stellar system?

In *Himalayas, Abode of Light* Roerich relates a conversation with a lama who put these questions: "Do you in the West know something about the Great Stone in which magic powers are concentrated? And do you know from which planet came this stone?" The passage clarifies the mystery— the astral stone must have come from another planet by a spaceship. The source of fallen meteorites cannot be identified.

Apollonius saw antigravity feats in Tibet, Saint-Germain

described a flight through space, Roerich spoke of a stone artifact from another solar system which he had carried on an expedition to Central Asia. There are invisible links connecting the emissaries of the City of Knowledge.

Historical records and legends disclose the fantastic fact that somewhere on this planet there might be a secret cosmic center of universal science with libraries, museums, and laboratories.

The life stories of Apollonius of Tyana, Saint-Germain, and Roerich serve as typical biographies of the men who have reached this fountain of primordial science in the past and have drunk of its waters.

22

In Quest of the Source

In this work an attempt has been made to show that there existed a much higher level of science and technology in antiquity and prehistory than is generally supposed.

Some of this knowledge is an enigma. For instance, how could the Mayas have devised a more precise calendar than we in this age of science? Why is the Khufu pyramid still the biggest megalithic edifice in the world? What made the Babylonians invent electric batteries 4,000 years ago? Why did the ancient Greeks and Romans expect to find planets beyond Saturn?

The scientist does not have a ready explanation for riddles of this kind because he is overburdened with the practical problems on hand. Moreover, the average man of science functions in a narrow field, possessing an in-depth knowledge of his branch of science but comparatively little about other branches outside of it. The scientists themselves admit this limited specialization and jokingly call it "professional cretinism." With the chain reaction in science and the overproduction of information, nothing can be done about this tendency for the moment. After all, the scientist is a man and not a computer.

The baffling aspects of the history of science are therefore left to theorists. However, true science is not merely the compiling of an immense catalogue of facts but also the ability to evaluate them. In order to throw some light on these confounding profiles of antique science, it is essential to formulate the problems.

—Why did there exist a long tradition of a Golden Age instead of that of a savage past?

—Why did knowledge rise and fall?

—What was it that prompted ancient Greek philosophers and Brahmin priests to discuss the possibility of inhabited worlds in space?

—Did the ancients possess the secret of antigravitation if they could raise heavy stones and even themselves into the air?

—Did the sages of former epochs know of energies other than electric power if they could construct perpetual lamps that burned for hundreds of years?

—How could the people of ancient times have left detailed descriptions of flying machines if such things had not existed?

—Were the automatons and robots recorded in mythology and history memories of automation in a past technological era?

—If civilized man is a recent arrival, who carved the images of animals extinct for hundreds of thousands of years?

—Can socialism be really new if the Incas had a moneyless economy with no private ownership?

—How could the *Vishnu Purana* give the geographical contours of the Americas and the North Pole zone without having made a scientific survey?

One way of resolving these awkward questions could be to visualize an advanced technological civilization which must have perished in a devastating geological cataclysm thousands of years ago.

Professor Frederick Soddy, a pioneer of atomics, wrote in 1909: "The legend of the Fall of Man, possibly, may be all that has survived of such a time before, for some unknown reason, the whole world was plunged back again under the undisputed sway of nature to begin once more its upward toilsome journey through the ages."

History, mythology, and the sacred books of most races support Professor Soddy's imaginative theory. Why did the legends of a deluge appear in sunny Egypt, Greece, and Mesopotamia; mountainous Peru and Mexico; icy Greenland; and the sandy Gobi?

The Cro-Magnon of 20,000 years ago was not physiologically any different from the average European. Had he been a survivor of another civilized epoch, we could then understand why he had the talent to produce works of art in his prehistoric museums of art, such as Altamira and Lascaux. In dynamic realism the masterpieces of this "caveman" were superior to the succeeding paintings and sculptures of Egypt, Babylon, or Crete.

In antiquity scholars believed in a lost legendary age. They were certain that empires had been destroyed by the wrath of the elements, and because of the planetary scope of this catastrophe, very little remained of their former grandeur except beautiful myths of a Golden Age.

For instance, Philo of Alexandria (20 B.C.-A.D. 54) wrote: "By reason of the constant and repeated destructions by water and fire, the later generations did not receive from the former the memory of the order and sequence of events."

In the *Timaeus* Plato (427-347 B.C.) recorded the words of an Egyptian priest: "There have been, and there will be again many destructions of mankind." When civilization is destroyed by natural calamities then "you have to begin all over again as children," said the high priest of Egypt to Solon.

The *Popol Vuh* of Guatemala laid stress on the great scientific knowledge of the "first men." "They were able to know all, and they examined the four corners, the four points of the arch of the sky and the round face of the earth." This primeval race could "see the large and the small in the sky and on earth," affirms the sacred book of the Mayas. But all their knowledge was lost when the gods asked: "Must they also be gods?" And so the "eyes of the first men were covered and they could see only what was close." This is how "the wisdom and all knowledge" of the first men were destroyed.

A myth is often history concealed. The story of the *Popul Vuh* may be true if a developed civilization had come to an end in a geological disaster and the survivors had been left with only a verbal tradition of a past epoch of culture.

Historical events settle down in time like mud in a river and are gradually covered up by new incidents. If the past is not uncovered, studied, and recorded, it will remain in oblivion. The amount of knowledge we possess about the past is only a small portion of the complete story of mankind.

There are riddles in the history of civilization which can be clarified if we take the legend of a sunken continent as a working hypothesis. In the last Ice Age, all of Canada, part of the United States, all of Belgium, Holland, Germany, and Scandinavia, and a portion of Eastern Europe were under an arctic ice sheet. About 12,000 years ago a sudden rise in temperature occurred, and the ice began to thaw. Sea levels rose 0.92 meters per century between 12,000 and 4000 B.C. during this deglaciation. What was the cause of the end of the Ice Age?

If the reality of Atlantis and its sinking are accepted, this could easily be the explanation of how the warm Gulf Stream began to flow northward when the barrier of the Atlantean continent was removed after its disappearance under water. In this manner, the stream became the "central heating" of Europe and warmed the climate so much that within 100 years giant masses of the European ice cap melted. If one were to put an island continent in the Atlantic to block the Gulf Stream, Europe as well as North America would again be transformed into an arctic zone tomorrow.

Global population figures between 6000 B.C. and the beginning of our era are extremely significant. There were about 250,000,000 people on earth 2,000 years ago. The population of the planet in 4800 B.C. was 20,000,000. In the year 5000 B.C. there were 10,000,000 on all the continents. One thousand years earlier—in 6000 B.C.—only 5,000,000 people inhabited the earth. On the basis of these figures, the population of the globe was well under 1,000,000 about 10,000 B.C.—an astonishingly low figure. Why was man such a rare creature if he has had a continuous existence as a primate and then as a rational being for at least 2,000,000 years? Was man with his works destroyed by the fury of the elements?

Now where should we turn to find the traces of antediluvian man? Egypt seems to be the first logical place to make a search for his records. "The Egyptians pride themselves on being the most ancient people in the world," said Pomponius Mela in the first century.

The story of Atlantis as told by Plato came to him from the priests of Neith in Sais through Solon. "I am what has been, what is, and what shall be," uttered the goddess to the priests, who kept chronicles of history for thousands of years. Herodotus admitted that he was unable to reveal certain mysteries which he had learned from the temple of Neith. These could have been revelations regarding the unknown history of mankind.

From Ammianus Marcellinus, a fourth-century Roman historian, to Ibn Abd Hokem, a ninth-century Arab savant, numerous records speak about pre-Deluge caches buried under or in the pyramids of Giza. Who knows whether these tales may have a grain of truth. The cosmic-ray probe of the pyramids by Ein Shams University of Cairo, initiated by Dr. Luis Alvarez, is in a position to provide interesting evidence in this field, particularly if the search is extended deep underground.

The X-raying of Khephren pyramid by cosmic rays in search of secret chambers has already disclosed an amazing phenomenon.* The tape-recordings for any one particular day differ markedly from those for the subsequent day. As cosmic rays shower upon the pyramid uniformly from all directions, the detector in the chamber at the bottom should show a uniform pattern. But if there were any vaults in the mass of the edifice above the detector, they would let more rays through than the solid areas and show as shadows on the detector. In September, 1968, the equipment was apparently in order, as the corners and sides of the Khephren pyramid could clearly be seen.

However, the hundreds of tins of recordings made by the scientists from 1967 to 1969 have disclosed a surprising fact. When they were put through the IBM 1130 computer at the Ein Shams University in Cairo, the pattern of daily recordings had no common features. Dr. Amr Gohed, who is in charge of the installation, remarked: "This is scientifically impossible. There is a mystery which is beyond explanation . . . there is some force that defies the laws of science at work in the pyramid." What force more potent than the cosmic rays did the builders of the pyramids possess? Had they left a machine radiating this power under the Giza pyramids?

In 1967 the author wrote an article for a Moscow newspaper entitled *Is There a Generator Under the Khufu Pyramid?* In alluding to the Sphinx he raised a question: "Is it really guarding the most ancient underground museum and library in the world?"

Whether evidence of an archaic civilization will ever be

*London *Times,* July 26, 1969.

found in Giza or in a submarine probe at the bottom of the Atlantic is still in the realm of speculation. But one thing is certain—the testimony of ancient writers, legends, and sacred books supports the reality of an advanced race which was not recorded in history. That nation could have passed its scientific tradition to the caste of priests of Egypt, Babylon, Mexico, and India and to the philosophers of Greece and China, which might explain how this knowledge continued throughout the ages.

The theory of an advanced culture and technology in protohistory which was devoured by a cataclysm can clarify many riddles in the history of science. For example, it can elucidate the following mysteries:

—Why the *Mahabharata* mentions planes and atomic bombs.
—Why ancient alchemists believed in the transmutation of the elements.
—Why the people of antiquity knew about Iceland, America, and Antarctica.
—Why the Mayan calendar is superior to ours.
—Why electric batteries were made in Babylon and India thousands of years ago.
—Why vaccination is described in the 3,500-year-old *Vedas*.
—Why the rock paintings of Altamira are such masterpieces of art.
—Why robots and computers are mentioned in classical as well as medieval writings.

But there is another hypothesis which can also explain a number of the enigmas posed in this book. It is fascinating yet utterly fantastic.

At a convention in New York the great physicist Niels Bohr once said to a scientist: "We all agree that your theory is mad.

The problem which divides us is this—is it sufficiently crazy to be right?"

Let the reader decide whether our conjecture is crazy enough to be sound. So we will begin at the beginning—the dawn of civilization. Strange demigods appeared on the world's scene who enlightened, taught, and helped primitive man.

A superior being once came to the land of the Nile in the distant past. He civilized the dwellers of Egypt by giving them symbols to record sounds and ideas, the harp to play upon, charts of the stars, numbers to count with, names of herbs, and remedies with which to cure sicknesses. Then the benefactor bid adieu to the people of Egypt and ascended into the sky. His name was Thoth, Hermes, or Mercury.

A culture bearer arrived in Greece in ancient times. He was a marvelous musician and possessed such wisdom that he could answer all questions. He spoke of such strange, incomprehensible things as, for instance, life in the stars. The Greeks knew him as Orpheus.

The Feathered Serpent or Quetzalcoatl descended from "a hole in the sky" when he came to Mexico. Another version describes a winged ship in which he sailed. Quetzalcoatl instructed the Central American Indians in the sciences of agriculture, astronomy, and architecture, and gave them a code of ethics. The civilizer left an indelible stamp on Central American culture and is still venerated in Mexico. Stained-glass windows and murals depicting Quetzalcoatl can be seen by tourists in the Palacio Nacional of Mexico City. Yet another legend about a cosmic torchbearer of civilization?

At the birth of Sumer, a weird creature landed on the shores of the Persian Gulf. It looked like a large fish but in its mouth there was a human face. This monster, which could have been a cosmic visitor in a space suit after the "splashdown," spoke to the primitive dwellers of Mesopotamia and taught them how to build cities, compile laws, plant wheat, write down thoughts, count with numbers, and observe the stars. This fishlike god, known as Oannes, civilized the savages and humanized their lives. His scientific legacy was of the highest quality, and the people of the Tigris and Euphrates valleys became great astronomers and mathematicians.

Far away from Babylon, in South America, a tall white man came from the land of dawn. He revealed to the Indians the secrets of civilization and inspired them with high ethical ideas. When his mission was completed, like Oannes, he disappeared into the sea. His name was Viracocha—"the foam of the sea." Yet another legend about a missionary from the stars, whose ship could sail on water as well as in space?

In recent years, in Tassili, northern Sahara, peculiar rock paintings were discovered. They depict men in what seem to be diving suits or space suits, without mouths. One six-meter rock image was aptly called the "Martian." Since the pictures are 9,000 years old, they are not likely to represent men with headgear protecting them against sand, as some skeptics have suggested, as at that time the northern Sahara was covered with rich vegetation!

This theory of the importation of science from a cosmic source is within the framework of scientific speculation.

A few years ago Frank Drake, the American astronomer, theorized that space visitors could have left caches of artifacts, tagged with radioactive isotopes. A scientifically mature humanity, reaching the level of the intelligence of cosmic travelers, would sooner or later spot these treasure vaults and discover souvenirs from a distant star. Perhaps the day has now arrived when we should test his theory by means of radiation detectors.

"Let the archaeologists take up the search for secret repositories established by cosmonauts. What if there were an immense treasure of the most precious scientific knowledge stored for us to inherit?" asks Dr. M. M. Agrest of the USSR.

Dr. Carl Sagan, an eminent astrophysicist in the United

States, speculates along similar lines: "The earth may have been visited many times by various galactic civilizations during geological times and it is not out of the question that artifacts of these visits still exist."

Professor Hermann Oberth, a pioneer in astronautics, stated a few years ago that visitors from space "have been examining the earth for centuries." He believes that an advanced galactic civilization could, by sharing its knowledge with us, raise our science to a level which would normally take 100,000 years to reach. Perhaps something of this sort has already taken place in the past history of humanity!

Interstellar travel seems to be a chimera because of the vast distances separating solar systems. To form an idea of the immensity of the universe, let us compare the sun to a billiard ball. Then the earth would be but a full stop on this page, placed 7.60 meters from the ball. The orbit of Pluto, which forms the borderline of our solar system on this scale, would be 300 meters from the sun. But the nearest star—Proxima Centauri—on the same scale would then be 2,000 kilometers away!

However, antimatter fuel and photon rays, as well as the shrinkage of time in a starship moving with a velocity close to that of light, could make interstellar travel a reality of the future. In fact, blueprints of these spaceships are already in existence, designed by the astronautical engineers of the United States and the USSR.

But what will be a technological achievement to us tomorrow may already be a means of transport to some other galactic civilization today. There is an eternity of time in the infinite universe. An older cosmic race could have mastered interstellar flight a million years ago and may even now be cruising in the expanses of our galaxy. They could have come here in bygone ages, and perhaps some folklore may refer to these visitations.

The view that contacts between worlds have taken place in

the past, are occurring at present, and will happen in the future is shared not only by science fiction writers but by some men of the highest caliber in academic circles.

In May, 1966, it was my privilege to meet in Nuremberg, Germany, the "father of modern rocketry," Professor Hermann Oberth. "I think over forty percent of the stars have planets and intelligent life might exist on some of them. To tell you the truth, this is the main reason why I became interested in astronautics in my early youth," said the distinguished scientist.

When I introduced the question of interstellar communication and possible surveillance of our planet by cosmonauts from another world in space, the professor's reaction was this: "It is the duty of the scientists to investigate this possibility."

In Paris in March, 1968, I had an interesting discussion with Dr. J. Allen Hynek, director of Dearborn Observatory and a professor at Northwestern University, who was the first in the world to employ satellites in astronomical observation.

"What percentage of solar systems could be suitable for life?" I asked.

"In my student days life in the universe was considered to be a freak. But now it turns out with modern theories of stellar evolution that, at least for a large class of stars, a planetary system may occur in the natural process of growth. To say that our star—the sun, is the only one to have planets will be akin to saying, for instance, that your cat can have kittens and no other cat in the world can have them. From the astronomer's viewpoint it is cosmically provincial to think that our solar system is the only one. There must exist around every star a 'temperate zone' in which conditions would be suitable to life."

"Is the possibility of visitations from superior galactic civilizations within the bounds of scientific speculation?"

"It is certainly within its bounds. As an astronomer I can

readily admit the strong possibility that there are other civilizations in our galaxy. But the engineering problems of getting here completely stagger my imagination. Nevertheless, I am perfectly willing to admit the reality of the ETI's [Extraterrestrial Intelligences]," responded Dr. Hynek.

During my stay in Moscow in November, 1966, I met Alexander Kazantsev, a famous science fiction writer. After World War II he gave a new interpretation to the Tunguska meteor catastrophe of 1908, which devastated a huge area in Siberia. He theorized that the explosion was caused by the detonation of a spaceship from another planet. This hypothesis received an active support from Boris Liapounov, a science writer of the USSR.

They speculated that a cylindrical body could not have been a meteor. Besides, it never reached the earth, as it blew up at an altitude of approximately 20,000 feet above the ground. Before the explosion the object performed an erratic maneuver changing its course. The permafrost, or the arctic soil frozen for thousands of years, was not pierced by the fragments of this "meteorite." According to witnesses the explosion was blinding even in daylight. Great devastation was brought about in the wild Siberian forest by it. When cut the trees showed an unusually thick ring for the year of 1908, suggesting radioactivity.

I visited Kazantsev in Moscow to get more facts about the Siberian mystery. While we were in the midst of an exciting conversation about the possibility that we earthmen might be the "grandsons of Mars," and that certain episodes from Biblical history could be records of the visits of cosmic guests to this planet, the doorbell rang. The host left the study, giving me an opportunity to make notes of this memorable dialogue.

In a few minutes Alexander Kazantsev returned with a booklet in his hand and a smile on his face. "We were right after all," he said, passing me a freshly printed brochure marked "The United Institute of Nuclear Research, Dubna. No. 6-3311." It contained a radiochemical analysis of tree ash from the site of the Tunguska explosion and alluded to the possibility of a contact of antimatter with matter.

One paragraph impressed me particularly: "In other words, we are coming back (no matter how fantastic it may seem) to the supposition that the Tunguska catastrophe was brought about by the crash-landing of a spaceship propelled by an antimatter fuel."

My friend Boris Liapounov showed me in Moscow a 1930 magazine, *Vestnik Znania,* which featured an article on life in other worlds. The great pioneer of astronautics Tsiolkovsky had this to say on the subject of interstellar communication: "If the machines of intelligent beings of other worlds have not visited the earth, it does not follow that they have not visited other planets. Secondly, to assess the fact of non-visitation of our world, we have at our disposal but a few thousand years of the conscious life of mankind. Think of the past and future ages!"

In the same journal Professor N. A. Rynin wrote that "If we turn to the tales and legends of hoary antiquity, we shall notice strange coincidences among legends in countries separated by oceans and deserts. This similarity of myths speaks for the visitation of earth by the inhabitants of other worlds in time immemorial."

These fantastic ideas have never been foreign to the author since his schooldays. If I am permitted to relate a personal experience from my early life in China, I will refer to a lecture given by me in Shanghai in 1941. In this address I spoke of "communication between planets by spaceships propelled by unknown energies," and the visits of superior intelligences to our planet earth. But my audience was not quite ready to accept the existence of beings in other cosmic worlds, and the lecture was not a great success.

But in our present era of space probes to Mars and Venus

and the voyages of earthmen to the moon, the public resistance to the concept of a plurality of inhabited planets has given wide cracks.

The problem of life in other stellar systems hinges on four basic questions:

—Is life on earth a unique phenomenon unparalleled in a universe with milliards of solar systems?
—Are we then alone in a dead universe?
—Is life potentially locked within the atom and therefore a universal attribute of matter?
—Are some worlds inhabited by rational beings at different stages of evolution?

Today in this age of space travel, there are but few people who would answer the first two questions in the affirmative.

Although the moon is desolate (nor does Mars look like a breeding ground for intelligent life), it is certain that somewhere in the limitless expanse of our galaxy with its 150 milliard suns, there are numerous planets similar to our earth which could have produced their own versions of life and consciousness.

To a cosmic tourist approaching the earth, our planet might give the impression of being uninhabited. In fact, that was the impression of Captain Lovell, an Apollo 8 astronaut, who broadcast this remark on his homeward flight from the moon: "What I keep imagining is that I am some lonely traveller from another planet. What would I think about the earth at this altitude [333,000 kilometers]? Whether I think it would be inhabited or not?"

The hypothesis of contacts of galactic civilizations with the men of earth in past ages can account for the following unexpected knowledge in remote antiquity:

—How the ancients were aware of the two moons of Mars.

—How the astronaut Hou Yih of China described the moon as "desolate, cold and glassy," forty-three centuries ago.

—How Greek philosophers were aware of the vast distances between the stars.

—Why myths about the descent of skybeings are global in scope.

—How antique thinkers were aware of planets beyond Saturn, which could not be seen without a telescope.

—How Sanskrit texts estimated the life-span of the universe in milliards of years, as does modern science.

—How the *Epic of Etana,* the *Book of the Dead,* and the *Book of Enoch* drew a picture of extraterrestrial space.

—How Saturn, although a dim stellar body, but actually the second largest planet in our system, was given the prominence it deserved in ancient pantheons.

Besides the importation of science from another stellar world or the existence of a cultural legacy from a vanished civilization, there is a less fantastic source of knowledge which could account for some of the peculiarities of the history of science.

In the course of the past 35,000 years man has been slowly reaching his present level. Since about 8000 B.C. he began to change the personality of a roving hunter for that of an established farmer and city tradesman.

The orthodox scientific view has no provision for an unknown advanced civilization in prehistory. It does not entertain ideas about culture bearers coming from cosmic space and accelerating the progress of mankind on this planet.

Anthropologists say that in the course of a long period of the development of the primates, a true man appeared. The Cro-Magnon prospered as a hunter and a fisherman. Members of a clan sat around the bonfire at night, listening to the elders relate their experiences, transmitting their accumulated knowledge to younger generations.

When man began to cultivate plants, domesticate animals,

212

and perfect his tools and weapons, he stepped over the threshold of civilization. That took place about 7,000 years ago.

Many of the accomplishments of humanity in former epochs which we outlined in this book can be explained by the march of progress.

But orthodox science is not in a position to offer an explanation to the unsolved mysteries of the history of science which have been examined in this work.

Now we come to the crucial point. Did the ancients receive a scientific legacy from the survivors of an older civilization destroyed by tidal waves and the fires of submarine volcanoes in a geological upheaval? Or was primordial science and culture brought to this world by the space visitors who had aeons ago reached the level of evolution on which we are standing now?

Actually there is no contradiction between the two hypotheses. As the great Tsiolkovsky said, our history is too brief to estimate how often visitations from space have occurred in the past. However, the American astrophysicist Carl Sagan believes that they take place at 5,500-year intervals.

If history stretches beyond the 7,000-year limit afforded to it by the historians, and if men lived in the last interglacial age, a descent of skybeings upon earth could have started an epoch of culture. If the technology of extraterrestrial intelligences was sufficiently developed to enable them to span astronomical distances, their way of life and thinking must have been equally advanced.

The starborn could have founded the world's first empires, ruled as the sun kings, and then passed on their authority to the so-called solar dynasties. This speculation is confirmed by a belief common to the legends of Egypt, India, China, Greece, Mexico, and Peru, which affirm that there was a time when the "gods" reigned over mankind.

This conviction is still alive in India and it has been my

experience to learn how strong it is. Upon my arrival a garland of tropical flowers was put on my neck by Indian friends who prostrated themselves at my feet to honor me as a "god." "We can't take any chances as you may be a skybeing from the stars merely pretending to have come from Australia," the Indians said, in response to my embarrassment and objections.

The principal thesis of this work, that the source of civilization lies farther back in time, will one day be vindicated. The origin of civilization is constantly receding as science advances. Since true science is fluid and modified by new evidence, it is not improbable that a great deal of the speculation offered in this study of the history of science will be found reasonable in the end.

Conclusion

In the nineteenth century a frozen Siberian mammoth was delivered to St. Petersburg for a demonstration at the session of the Imperial Academy of Sciences. A lecture on the prehistoric beast was given by a member of the academy and then the savants were invited to a banquet.

After the dinner the president of the Academy of Sciences announced: "Gentlemen, you have just eaten the oldest steak in the world—it came from our Siberian mammoth!" There were smiles as well as green faces in that dining room.

This book is not unlike that prehistoric dinner. Finding computers, aircraft, electricity, penicillin, robots, atomic theory, and descriptions of space and the moon in hoary antiquity is like having a mammoth steak on one's plate. But it was still fresh and tasty, although some 12,000 years old. The tales of ancient times told here are equally fresh today. They link the past with the present.

The analogies of antique scientific notions with our present-day technological and scientific attainments can hardly be contested. This is the midday of technology but there have been brilliant days of theoretical science in the past.

We owe much more to our predecessors than we realize, if not for the actual technical devices and scientific knowledge we have today, then at least for their concepts. Lucian anticipated Apollo probes. Democritus worked out the basis of atomic theory. Ancient alchemists forestalled modern chemical transmutation of the elements. Wonder drugs can be traced to ancient Egypt. Aristarchus of Samos promulgated the heliocentric theory eighteen centuries before Copernicus. The Sanskrit *Book of Manu* proposed the idea of evolution long before Darwin. Seneca in *Medea* disclosed his knowledge of North America. The Babylonians had an electric battery 3,800 years before Volta. The Indian classics described aircraft in great detail. The ancient Greeks had robots and computers centuries before this age of automation. The list is long and our debt to antiquity and prehistory is great.

But we are in even greater debt to those mysterious torchbearers of civilization who, at the dawn of history, imparted their knowledge to the astronomer-priests all over the globe.

There was a golden age in which the miracles of science were as commonplace as they are now. The source of this forgotten science must be sought in time as well as in space.

Rediscovery of Science

Scientific and Technological Ideas	Known in Antiquity	Rediscovered
Atomic theory	Uluka Kanada (c. 500 B.C.) Democritus (460-361 B.C.) Leucippus (b. c. 480 B.C.) Epicurus (341-270 B.C.)	Boyle (1661) Dalton (1805)
Theory of Relativity	Heraclitus (c. 540-475 B.C.) Zeno of Elea (5th century B.C.)	Einstein (1916)
Transmutation	The alchemists (1st century B.C.—1st A.D.)	Rutherford (1919)
Age of the earth	Life-span of the universe 4.32 billion years (the *Mahabharata* and the *Puranas*)	4.6 billion years (20th century)
Formation of the solar system	The *Popul Vuh*, Wang Chung (A.D. 82)	Kant (1755) Laplace (1796)
Evolution	Anaximander (c. 611-547 B.C.) Book of Manu (c. 200 B.C.)	Darwin (1859)
Earth as a planet	Pythagoras (6th century B.C.) Anaximander (611-547 B.C.) Heracleides of Pontus (388-315 B.C.)	Copernicus (1473-1543)
The moon shining by reflected light	Parmenides (b. c. 544 B.C.) Plutarch (1st century A.D.)	Galileo (1610)
The moon's connection with tides	Posidonius (135-50 B.C.)	Kepler (1571-1630)
Planets beyond Saturn	Democritus (5th century B.C.) Anaximenes (5th century B.C.) Seneca (1st century A.D.)	Uranus (1781) Neptune (1846) Pluto (1930)
Sunspots	Chinese astronomers 2,000 years ago	Galileo (1610)
Jupiter's four largest moons, phases of Venus, seven satellites of Saturn	Babylonian priests (c. 2000 B.C.)	Galileo (1610) Cassini, Huygens, Herschel, Bond (17th-19th century)

217

Scientific and Technological Ideas	Known in Antiquity	Rediscovered
Milky Way—a cloud of stars	Democritus (5th century B.C.)	Galileo (1610)
Meteorites—stones from space	Diogenes of Apollonia (5th century B.C.)	Academy of Sciences, Paris (1803)
Extraterrestrial space	*The Epic of Etana* (2700 B.C.), *The Book of the Dead* (*c.* 1500 B.C.), The Book of Enoch (2nd century B.C.)	Gagarin (1961)
Music of the Spheres	Pythagoras (6th century B.C.)	Radio astronomers Jansky-Reber (1930s 1940s)
Electric batteries	Babylon batteries (2,000 years old)	Volta (1800)
Aviation	Daedalus (2500 B.C.), Emperor Shun (2258-2208 B.C.), Ki Kung Shi (1766 B.C.), etc.	Wright brothers (1903)
Turbo-jet engine	Heron (1st century B.C.)	Von Ohain (1939) Whittle (1941)
Space travel	Orbiting the earth (the *Surya Siddhanta*, 2,000 years old)	Sputnik I (1957)
Robots and computers	Daedalus' automatons (2500 B.C.), Antikythera computer (65 B.C.), etc.	Wiener (1950s)
Plumbing and sanitation	Knossos (2000 B.C.), Mohenjo Daro, Harappa (2500 B.C.)	19th century
City planning	Mohenjo Daro, Harappa (2500 B.C.)	Paris, Washington (17th-18th centuries)
Vaccination	The *Vedas* (1500 B.C.)	Jenner (1749-1823)
Penicillin	Egypt (*c.* 2000 B.C.)	Fleming (1928)
Existence of America	Plato (4th century B.C.) Seneca (1st century A.D.) The *Vishnu Purana* (200 A.D.)	Bjarni and Ericson (*c.* 1000) Columbus (1492)

Bibliography

Braun, W., and Ordway, F. I., *History of Rocketry and Space Travel*. Thomas Y. Crowell, New York, 1966.

Brewer, E. C., *A Dictionary of Miracles*. Lippincott, Philadelphia.

Burgess, E., *Surya Siddhanta*. New York, 1860.

Clagett, M., *Greek Science in Antiquity*. Abelard-Schuman, New York, 1955.

Cohen, J., *Human Robots in Myth and Science*. George Allen & Unwin, London, 1966.

Cooper-Oakley, I., *The Comte de St. Germain*. Ars Regia, Milan, 1912.

Cottrell, L., *Wonders of Antiquity*. Longmans, London, 1959.

Dikshitar, R. Ramachandra, *Warfare in Ancient India*. Macmillan, 1944.

Doberer, K. K., *The Goldmakers*. Nicholson & Watson, London, 1948.

Draper, J. W., *History of the Conflict Between Science and Religion*. London, 1885.

Duhem, J., *Histoire des Idées Aeronautiques*. Paris, 1943.

Figuier, L., *L'Alchimie et Les Alchimistes*. Hachette, Paris, 1860.

Frazer, J. G., *The Golden Bough*. Macmillan, New York, 1937.

Giles, L., *A Gallery of Chinese Immortals*. John Murray, London, 1948.

von Hagen, V. W., *Highway of the Sun*. Duell, Sloan & Pearce, New York, 1955.

————, *Realm of the Incas*. Mentor, New York, 1957.

Hall, M. P., *Masonic, Hermetic, Qabbalistic and Rosicrucian Symbolical Philosophy*. P.R.S., Los Angeles, 1947.

Hapgood, C. H., *Maps of the Ancient Sea Kings*. Chilton Books, New York, 1966.

Hawkins, G. S., *Stonehenge Decoded*. Souvenir Press, London, 1965.

Kondratov, A. M., *Lost Civilizations*. (Russ.), Moscow, 1968.

Larguier, L., *La Faiseur d'Or*. J'ai Lu, Paris, 1969.

Larousse, *Encyclopedia of Mythology*. Paul Hamlyn, London, 1960.

Laufer, B., *Prehistory of Aviation*. Field Museum, Chicago, 1928.

Ley, W., *Watchers of the Skies*. Viking Press, New York, 1963.

Magre, M., *The Return of the Magi*. Philip Allan, London, 1931.

Mason, S. F., *Main Currents of Scientific Thought*. Abelard-Schuman, New York, 1956.

Mehta, C. N., *The Flight of Hanuman to Lanka*. Narayan Niketan, Bombay, 1940.

Mellersh, H. E. L., *From Ape Man to Homer*. Scientific Book Club, London, 1962.

Metraux, G. S., and Crouzet, F., *The Evolution of Science*. Mentor, New York, 1963.

Moura, J., and Louvet, P., *Saint Germain*. J'ai Lu, Paris, 1969.

Neugebauer, O., *The Exact Sciences in Antiquity*. Ejnar Munksgaard, Copenhagen, 1951.

Philostratus, *Life and Times of Apollonius of Tyana*. Stanford University, 1923.

The Popul Vuh. Wm. Hodge, London, 1951.

Redgrove, H. S., *Alchemy—Ancient and Modern*. Rider, London, 1922.

Roerich, N., *Heart of Asia*. Roerich Museum Press, New York, 1929.

———, *Himalaya—Abode of Light*. Nolanda, Bombay, 1947.

Sainte-Hilaire, J., *On Eastern Crossroads*. Frederick A. Stokes, New York, 1930.

de Santillana, G., *The Origins of Scientific Thought*. Weidenfeld & Nicolson, London, 1961.

Sarton, G., *Ancient Science and Modern Civilization*. University of Nebraska, Lincoln, 1954.

Schwartz, G., and Bishop, P. W., *The Origins of Science*. Basic Books, New York, 1958.

Seal, B., *Positive Sciences of the Ancient Hindus*. Longmans Green, London, 1915.

Sedgwick, W. T., and Tyler, H. W., *A Short History of Science*. Macmillan, New York, 1952.

Shklovsky, I. S., *Life and Intelligence in the Universe*. Moscow, 1965.

Siegel, F. J., *Life in the Universe*. (Russ.), Minsk, 1966.

Singer, C., *A Short History of Scientific Ideas*. Oxford, 1959.

Soddy, F., *The Interpretation of Radium*. John Murray, London, 1909.

Sullivan, W., *We Are Not Alone*. Hodder & Stoughton, London, 1965.

Taylor, F. S., *The Alchemists*. Henry Schuman, New York, 1949.

Thomas, P., *Epics, Myths and Legends of India.* D. B. Taraporevalla, Bombay, 1961.

Werner, E. T. C., *Myths and Legends of China.* George G. Harrap, London, 1922.